Spanning the Gap: Unveiling the Secrets of Tied-Arch Bridges

Johnny

TABLE OF CONTENTS

1. INTRODUCTION

1.1. Background

Arch bridges have been built for many centuries, using construction materials such as stone masonry and timber. The primary load transfer mechanism of arch bridges is compression and this makes the structural form attractive for materials with high compressive strength (Figure 1). The tied-arch bridge, which is a contemporary variant of the generic arch bridge, suspends the road way below the arch ribs and is often used when there is free space below the bridge deck or when the conventional arch founding conditions are not available (Stavridis, 2010). Modern concrete and steel materials have allowed engineers to build longer span bridges with more slender sections than had been done in the past.

Figure 1 Salginatobel bridge, designed by Robert Maillart (1929- 1930)

The tied-arch bridge is commonly constructed with structural steel, because of the inherently high tensile and compressive capacity of steel. Concrete on the other hand is known to have excellent compressive capabilities, but it is also known to be equally poor in tension. The advent of prestressing technology in the early twentieth century has however opened avenues of new application for conventional concrete by improving the tensile capacity of the composite concrete material. Therefore, designers have the opportunity to design tied-arch bridges with either the structural steel or prestressed concrete material.

In South Africa the most commonly used material for the construction of small to medium span freeway bridges is concrete. Typically small spans range between 20 and 50 m, whereas medium spans range between 50 and 80 m. On the Western Cape provincial freeways only 3 of the 104 bridges are constructed with steel girders and the trend is similar for the rest of the country (Haas, 2014). The main factors that influence the material choice for bridge girders are:

- Bridge span
- Location
- Experience of the bridge engineer working with structural steel or concrete;
- Material cost and availability
- Available construction technology
- Time constraints
- Safety requirements

During a survey conducted among South African bridge design professionals, there was an 80% level of agreement that steel is too expensive and that maintenance costs associated with steel construction are too high (Haas, 2014). Therefore it is understandable that bridges in South Africa are predominantly made of concrete.

To achieve longer spans with the concrete material, the deck is usually prestressed with draped cables which counteract the heavy weight of the deck. The prestressing technique is a specialised method which allows the high tensile strength of the prestressing steel to be utilised to enhance the performance of concrete, both for the ultimate limit state and the serviceability limit state of the structure.

The deck of a concrete tied-arch bridge is also prestressed, but not for the conventional reason as mentioned above, but rather for countering the thrust action of the arch. This is achieved by placing concentric tendons in the tie-beam.

Design requirements for prestressed bridges are usually governed by the serviceability limit state of cracking (Robberts & Marshall, 2000). To this end, it is often a client requirement that prestressed structures are crack-free during construction and during its service life. To achieve this, prestressed concrete members have to be evaluated at various critical stages in the life of the structure, when cracking may take place. The stages immediately after post-tensioning and when the structure is subject to the most onerous traffic load at the end of its lifespan has been identified as being the most important design scenarios to analyse (Robberts & Marshall, 2000).

If the cracking limit state is exceeded, it is important that the anticipated behaviour of the structure is understood and that it does not adversely affect the structure. Cracking in the structure would influence the stiffness and durability of the structure and is therefore undesirable.

This book is based on the design and construction of Ashton Arch Bridge (Figure 2), located approximately 150 kilometres north of Cape Town, South Africa. To the knowledge of the author, this signature structure will be the second prestressed concrete tied arch-bridge in South Africa, after the Nico Malan Bridge (Huisman, 1972). The new bridge consists of a cable-supported concrete deck which spans 110 metres between abutments. The arch ribs rise approximately 23 metres above deck level and is restrained by two parallel prestressed concrete tie-members.

Figure 2 Ashton Arch Bridge during construction. Photographed by Terry February for AECOM, 2018.

For this structure, additional precautions must be taken to ensure that the non-redundant members, such as the tie-beam and arch ribs, function as the designer intended (further detail provided in Section 2.2). It was therefore deemed necessary to verify the behaviour of these critical members.

1.2. Problem statement

The construction of complex bridges often takes several years and in this time the bridge's structural system continually changes. When developing a finite element method (FEM) model, the construction stages that alter the stress state of the structure must be identified and modelled. A static model of the partially completed bridge must then be analysed at each construction stage to determine the structural behaviour at a particular time. As the bridge is constructed, the stress state of the various members continually changes until it is complete and thereafter the stress state further changes due to creep and shrinkage effects, until some future design time. The final bridge structural system may then be said to have built in stresses, and this final state may then be used to design the structure with the maximum service loads applied to it.

The major construction events wherein the stress state of the bridge changes are:

- Post-tensioning of Tie-beams beams

4

- Post-tensioning of Transverse beams

- Post-tensioning of secondary Longitudinal beams

- Completion of both concrete arch ribs and release of supporting falsework

- Installation and tensioning of cable hangers

During these stages the structure should be checked to ensure that structural elements do not exceed serviceability limits and that the structure is constructed safely. FEM software packages were used to analyse each construction stage and the in-service behaviour of the concrete tied-arch bridge (Figure 3). The following key observations were made when reviewing the analysis:

- Although the critical members remain in compression for the majority of its design life, the model indicated that there is a possibility of tension fibre stresses developing in the Tie-beam members under the maximum service load

- The analysis indicated that there are vulnerable zones on the Tie-beams, where the members are susceptible to tension fibre stresses

- It was however noticed that these stresses were not excessive and could be managed by altering the construction methodology

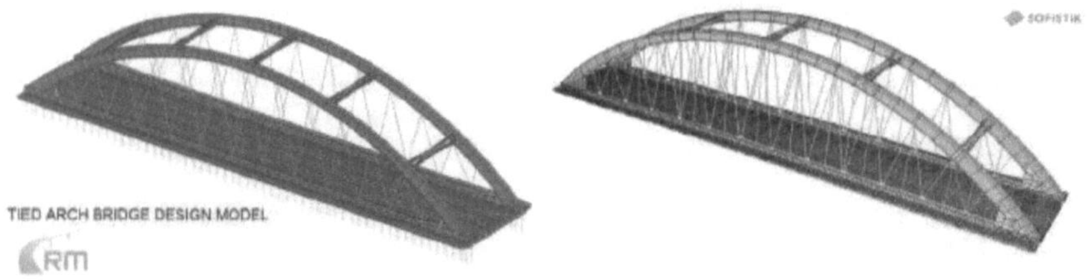

Bentley RM Bridge   SOFiSTiK FEM Model

Figure 3 Ashton Arch Bridge FEM Models

Since the tie-beam and arch rib members are critical to the overall performance of the structure, it was decided to monitor their behaviour during erection, to ensure that the structural behaviour of the FEM model correlates with the actual structure. The prestress behaviour of tie-beam members and state of strain at vulnerable zones must be clearly understood to ensure that cracking and reduced serviceability is avoided.

1.3. Scope and Objective

The *general objective* of the research presented in this book is to validate the structural behaviour of the tie-beams, after completion of the post-tensioning activity. This objective will be achieved by validating model parameters with field conducted tests performed during post-tensioning.

The *Input* parameter, which is the prestress loading onto the structure, will be validated with:

- tendon elongation measurement during post-tensioning
- tendon lift-off tests to determine force transmission between the near and far anchors

The *Output* parameter, which is the strain and displacement response, will be validated by:

- measuring the deck deformation before and after post-tensioning
- strain measurement at cross-sections identified as a vulnerable zones

Validation of the different parameters are accomplished by comparing the results of theoretical models with data received from the field-testing during construction of the bridge. The different field tests may be regarded as the *specific objectives* of this research project, which when evaluated holistically will validate the *general objective*.

If the structural behaviour at vulnerable zones in the tie-beam can be validated after post-tensioning, then the change in the stress state, as a result of the remaining construction stages may be summed to it by the theory of superposition. The most onerous in-service loads may also be further superpositioned to determine the maximum fibre stresses experienced during the life-span of the structure. This procedure will provide assurance that tensile fibre stresses of critical members remain within design limits throughout the design life of the structure. The concept of superposition of loads onto a tied-arch bridge is demonstrated in Figure 4.

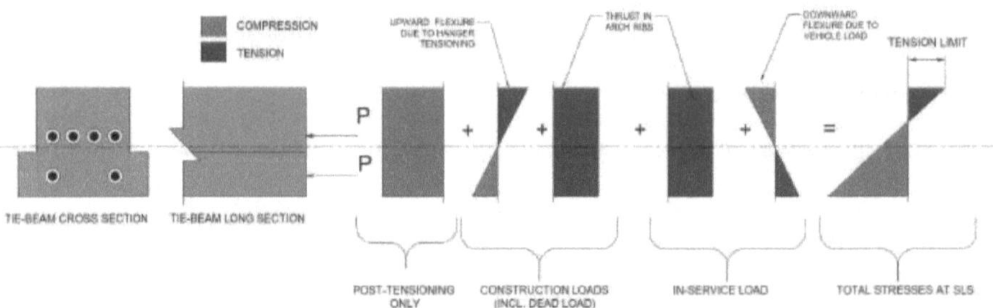

Figure 4 Superposition of loads

Although the evaluation of the maximum fibre stress is a project requirement for the construction of the Ashton arch bridge, it will not be checked as part of this book. This book also excludes *model updating* required to reconcile differences encountered during validation.

Additional measurements will be done in other critical elements, such as the arch ribs, throughout the construction of the bridge but these tests are beyond the scope of this book. The scope is limited to measurements taken during the post-tensioning of the deck.

2. REVIEW OF THE TIED-ARCH BRIDGE FORM & BEHAVIOUR

2.1. Types of arch bridges

Arch bridges may be classified according to the following criteria:

a) Material

b) Structural articulation

c) Shape of the arch

In recent times the construction material of choice has either been reinforced concrete or structural steel. Articulation refers to the degree of fixities at the support, which could be fixed or hinged. Generally fixed supports are preferred for concrete structures, whilst fixed or hinged conditions are common for steel structures. Fixed arches are generally designed with steep curves, while hinged arches may be flatter (Xanthakos, 1994).

The optimum shape of an arch is a catenary, but this is very closely mimicked by the more familiar parabola shape. Other modern shapes include elliptical, semi-circular or a compound circular curve (Xanthakos, 1994). Segmental arch shapes, like the Wild Bridge in Voelkermarkt, Austria are also an attractive modern option, because of the possibility of utilising the pre-casting construction method. These alternative shapes will attract greater bending moments in the arch, but may be accounted for when sizing the members in design.

Figure 5 Arch shapes

Arch bridges may be further grouped into three different types, which are differentiated according to the position of the deck relative to the arch.

The *Deck type arch* bridge, shown in Figure 6 a, is the most common configuration and is arranged so that the arch is below the bridge deck. This type of arrangement is also known as the true arch or the perfect arch, because the load applied to the arch is primarily uniformly distributed and therefore the arch acts almost purely in compression. In the Roman and Renaissance periods these arch barrels would have been built with wedges of tapered-stone masonry which is different to the monolithic steel or concrete materials used in more recent times. The engineering properties of monolithic materials, such as the thermal and shrinkage behaviour, are better understood by engineers and therefore longer span structures can be built with these materials (Xanthakos, 1994).

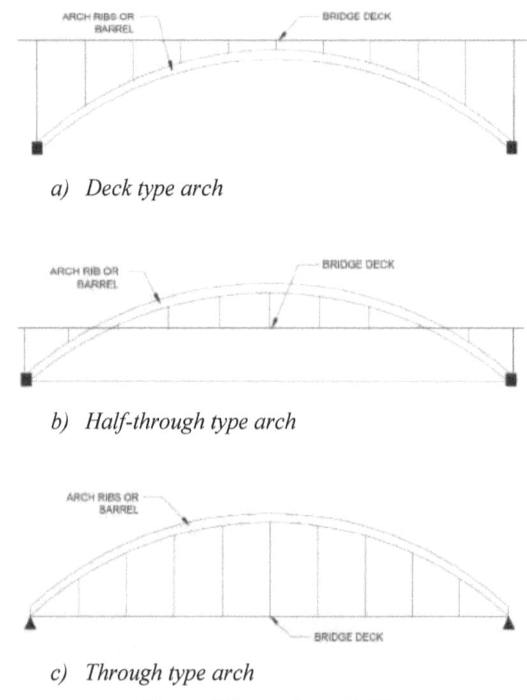

a) *Deck type arch*

b) *Half-through type arch*

c) *Through type arch*

Figure 6 Types of arch bridges

If the *Deck type arch* is loaded with fill material, then the bridge is known as a *solid spandrel deck arch*. If the space between the arch and the deck is open or has columns supporting the deck, the bridge is known as an *open spandrel deck arch*. The South African N2 garden route has three significant open spandrel deck type arch bridges, which were built between 1950 and 1985. The three bridges are shown in Figure 7, with the Bloukrans bridge being the most recent and is considered the last mega-bridge constructed in South Africa

a) Paul Sauer Bridge (1956)   b) Van Stadens Bridge (1971)   c) Bloukrans Bridge (1983)

Figure 7 South African Deck type arch bridges

Two other varieties of the arch bridge are also noteworthy. This is the *Half-through type arch* bridge and the *Through type arch* bridge shown in Figure 6 b and c. For these types of arch bridges the deck is located below the crown of arch ribs or barrel. The deck is often supported from the arch by hangers, which act primarily in tension. The difference between these two systems is that for the *Half-through arch* the deck is located at an elevation in-between the arch rib crest and the spring point of the arch rib. For the *Through type arch* the deck is located at the same elevation as the arch spring point. South African examples of these two variants are shown in Figure 8, which is the Umtamvuna River Bridge in Port Edward and the Nico Malan Bridge located in Port Alfred.

<table>
<tr><td>a) Umtamvuna River Bridge (1950)
Half-through type arch</td><td>b) Nico Malan Bridge (1970)
Through type arch</td></tr>
</table>

Figure 8 Half-through and Through type arch bridges

Conventional arch bridges resist the thrust of the arch by an external boundary, such as natural rock foundations or large concrete foundations. A *Tied-arch bridge* counteracts the thrust of the arch by using a tensioning member located at the level of the bridge deck. *Tied-arch* bridges may therefore be a *Half-through type arch* or a *Through type arch* bridge. It is important to note that the Through type and Half-though type arch bridges are not always tied-arch structures, since the lateral loads may be kept in equilibrium by other means. 'Tied-arch' is synonymous with 'bow-string' arch and the nomenclature can be used interchangeably.

Free body diagrams of *tied-arch* systems are shown in Figure 9 a-c.

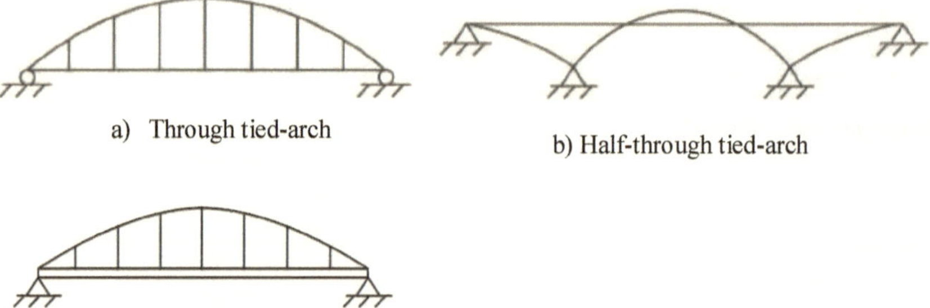

<table>
<tr><td>a) Through tied-arch</td><td>b) Half-through tied-arch</td></tr>
</table>

c) Through tied-arch with stiffened deck

Figure 9 Types of tied-arch bridges (Xanthakos, 1994)

The *tied-arch* may either have a slender tie with a deep stiff arch rib or the opposite arrangement as shown in Figure 9 c (Chen & Duan, 2000). This is because transient loads will result in non-zero bending moments in the arch ribs and therefore it is often necessary to design arch bridges with moment resisting ribs. Alternatively the deck may be stiffened, so as to transfer a more uniform load to the arch and therefore the arch rib may be designed as a more slender element (Xanthakos, 1994).

2.2. Tied-arch bridges

The primary members of a tied-arch bridge are the arching ribs and the tie-beam members. The arch ribs are integrally connected to the tie-beam at the spring point, but are also connected throughout the length of the deck by hanger cables, with an articulated connection. The hanger cables are spaced evenly to ensure uniform loading of the arch ribs by the self-weight of the bridge deck. The nomenclature and force distribution of a typical tied-arch bridge is illustrated in Figure 10.

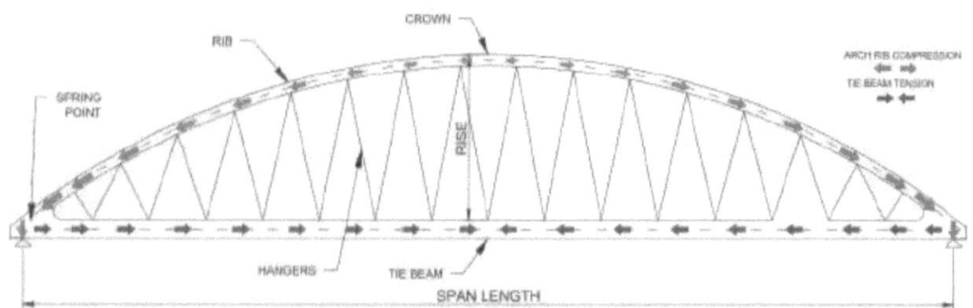

Figure 10 Force flow in a tied-arch bridge

Arch theory implies that a parabolic arch rib loaded with a uniformly distributed load will transfer the forces purely via axial thrust to the supports. The thrust exerted by the arch ribs are restrained by structural "tie" members connecting the ends of the arch rib, and therefore bridges of this kind are known as tied-arch bridges. The force distribution in the structure may be regarded as self-contained, and the only load transferred to external bearings are the vertical reactions of the structure. The structure is therefore regarded as simply supported. The transfer of load is illustrated by the four steps shown in Figure 11.

1 Self-weight of the bridge deck is applied to the structural system as a uniformly distributed load (UDL)

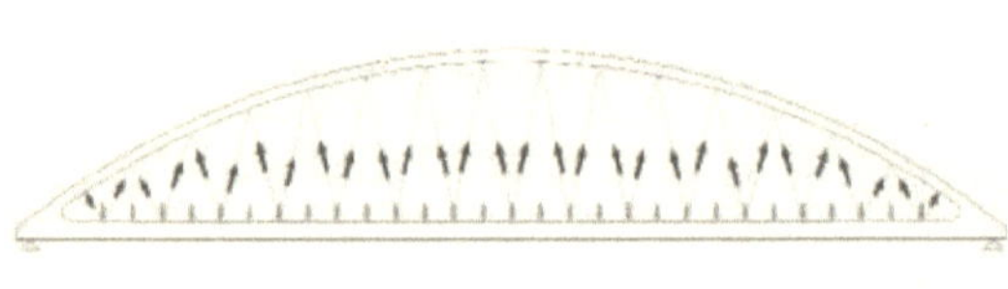

2 The load is transferred to the hangers, which are taut between the arch rib and the tie-beam

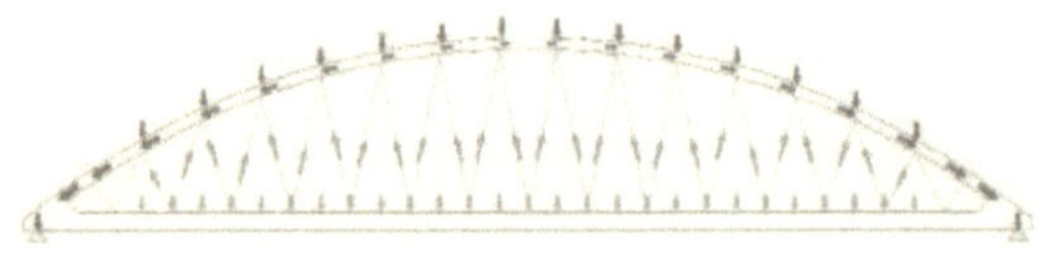

3 The hangers transfer the load to the arch ribs at discrete locations, but the global effect may be regarded as a UDL. The arch resists this load in compression

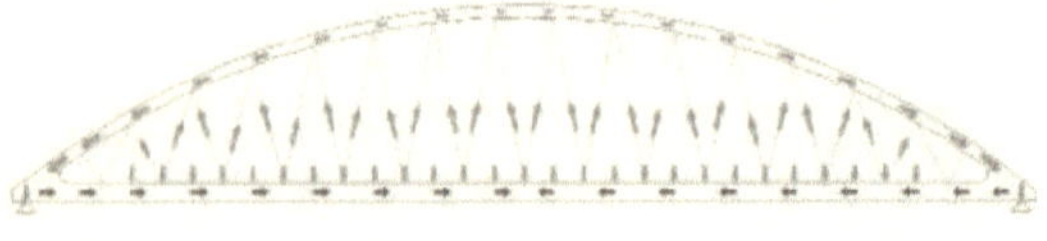

4 The axial thrust of the arch is transferred to the tie-beams, which restrains the outward thrust and keeps the system in equilibrium by acting in tension

Figure 11 Load transfer through tied-arch system

The tie-beam members are designed to work in tension by virtue of their structural arrangement. In addition to counteracting the thrust in tension, the tie-members also carry the load of the deck in flexure, between hanger connection points. The tie-beam member may therefore also be thought of as a continuously supported beam.

Examples of Steel *Tied-arch* bridges can be found effortlessly, but concrete *Tied-arch* bridges are rare. According to an article by Weaver (2011), there were only two modern concrete tied-arch bridges in the United Stated in 2012. The Covered Bridge in Maine, completed in 2012 and the Depot Street Bridge in Oregon, completed in 2006. The Covered Bridge has two parallel post-tensioned concrete tie-girders which span 91.5 m and two parallel arch ribs with a rise of 18 m. The Depot Street bridge is a 93-metre-long bridge and was constructed along side the existing bridge before being pushed laterally into its final position on completion of demolition of the existing structure. South Africa boasts its own post-tensioned concrete tied-arch bridge in Port Alfred, Eastern Cape, which was opened to traffic in December 1971. The Nico Malan bridge has a clear span of 84 m with reinforced concrete ribs which rise to a height of over 17 metres. The tie is connected to the arch ribs with vertical prestressed concrete hangers, spaced 5.3 metre apart (Huisman, 1972). Photographs of the three bridges are shown in Figure 12.

a) Depot Street bridge (2006) b) Covered Bridge (2012)   c) Nico Malan bridge (1971)

Figure 12 Concrete Tied-arch bridges

High strength 50 MPa concrete has been specified for the main arch rib and tie-beam members of the Ashton Arch bridge. It is however known that concrete has a low tensile capacity and therefore it is necessary that the tie beam elements be prestressed so that it can carry the outward thrust applied by the arch ribs. This is a common feature for concrete tied-arch bridges as is shown in Figure 13.

a) Covered Bridge   b) Ashton Arch Bridge

Figure 13 Post-tensioning of concrete tied-arch bridges

A tied-arch bridge may also be classified according to the arrangement of the hangers. The conventional hanger arrangements are vertically orientated. Nielsen then patented a V-configuration of the hangers in 1926 and defined that each hanger may cross each other not more than once. This arrangement was adopted for the Ashton arch bridge. The V-configuration provides a dramatic improvement on the flexural performance of the deck and arch ribs. A drawback on this method was that compressive forces may arise in the inclined hangers if the live load to dead load ratio was too high, as is evident with light weight bridges. Later in the 1950's, Dr. Tveit developed a hanger arrangement where each hanger cross each other at least twice, known as the *network ach* (Mato, Cornejo, & Sánchez, 2011). This arrangement, with multiple hanger intersections, resulted in a noticeable reduction in the possibility of the hangers going into compression under transient loads (Mato, Cornejo, & Sánchez, 2011). Different hanger arrangements are shown in Figure 14.

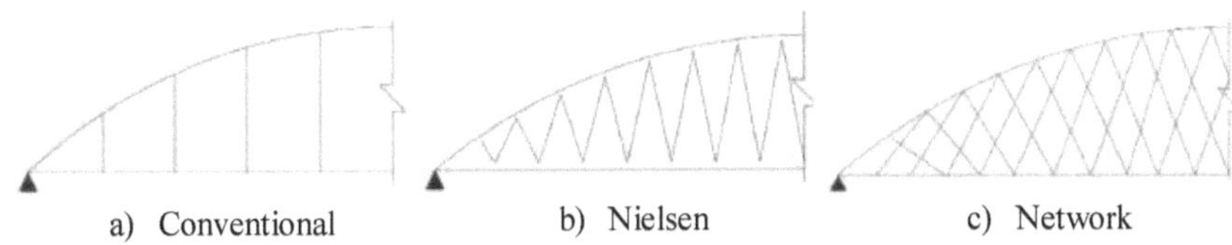

Figure 14 Network type tied-arch

Although the structural form of a concrete tied-arch bridge is elegant because of its ability to express symmetry, regularity and visualisation of force flow, it is often ignored as the material of choice for tied-arches. Hewson (2012) states that "lighter steel or composite decks are usually preferred over heavier concrete arrangements" for arch bridges when the deck is suspended from the arch ribs by means of cables. Stavridis (2010) states that for a tied-arch system the deck may be made out of prestressed concrete, but that it is more suitable for design and aesthetics that the deck is made of steel (Stavridis, 2010, p. 311).

2.3. Design considerations

When considering the *conceptual design*, a deck type arch bridges require favourable topography, such as large V-shaped gorges, while tied- arch structures can be built on foundations that are similar for conventional bridges. The typical span length of arch bridges range between 50 metres and 200 metres and their span to rise ratio is dependent on the articulation at the support. The span to rise ratio of the Ashton tied-arch bridge is 5:1, which is close to the upper bound for fixed articulation arch structures in Table 1.

Table 1 Typical Span to Rise Ratios for arch bridges

Articulation	Lower bound	Upper bound	
Fixed arch (steep)	3: 1	7: 1	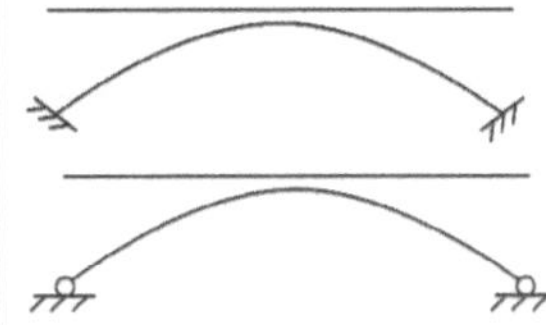
Two-hinge arch (flat)	6: 1	10: 1	

16

For the *detailed design* of a concrete structure, the designer should be cognisant that the aim of design is to achieve a structure which is safe, serviceable, durable, economical and aesthetically pleasing (Robberts & Marshall, 2000). These requirements are addressed by designing according to the *Limit States Design Method.*

2.3.1. Limit states Design

The Limit State Design Method is prescribed for the design of highway bridges and culverts by TMH7: 1981 Parts 1 and 2. The procedure is intended to ensure that the structure being designed remains fit for the required purpose during the expected design life (Committee of State Road Authorities, 1981).

A structure may become unfit for use when it exceeds a particular state, called a limit state. These limit states are categorised as the *Ultimate Limit State* and the *Serviceability Limit State.* The *Ultimate Limit State* corresponds to the maximum load carrying capacity of a structure and the *Serviceability Limit State* relates to criteria governing normal use and durability (Committee of State Road Authorities, 1981).

For a particular limit state, the designer will ensure that the actions applied to a structure does not exceed the resistance of the structure's members. The actions applied to the structure is variable in nature and therefore they should incorporate the appropriate partial factor to allow for their uncertainty. Similarly, the resistance of a structural member has variable material properties which is also accounted for by some partial safety factor. These partial factors may be probabilistically or empirically determined and accounts for the consequence of failure as well (Committee of State Road Authorities, 1981).

2.3.2. Preliminary arch analysis

When commencing the detailed design of an arch bridge, preliminary hand calculations can be made with a uniformly distributed load applied to the arch ribs. The vertical reaction forces can be determined by equilibrium of forces in the vertical direction in Figure 15 (a) and Equation 1. The horizontal thrust restraining the arch rib can be calculated by taking moments about the arch crown and equating the formulae to zero as shown in Figure 15 (b). This restraining force may represent a tie-beam or an external horizontal support. The optimum parabolic shape of an arch which does not induce any bending moments can be calculated using the horizontal thrust calculated in Equation 2 and 3 (Reis & Pedro, 2019).

$$V = {qL}/{2} \tag{1}$$

$$H = {qL^2}/{8f} \tag{2}$$

$$y(x) = -\frac{q}{2H}x^2 + \frac{qL}{2H}x \tag{3}$$

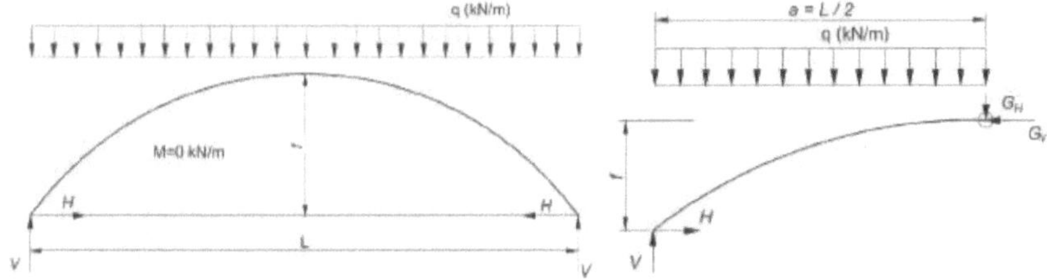

a)   UDL applied to elastic arch   b)   Derivation of horizontal thrust *H*

Figure 15 Preliminary arch analysis

The symbols referred to in equations 1, 2 and 3 and Figure 15 are as follows:

$x \& y$ - points of interest on the arch rib L – Span Length

f - height of the arch crown above support M – Bending Moment

H - Horizontal Reaction at bearing pt Q – load per unit length

V - Vertical Reaction at Bearing Pt

2.3.3. Stability of Compression-Members

Since the arch ribs are compression members, they may be susceptible to in-plane buckling at the ultimate limit state. An eigenvalue analysis is therefore necessary to determine the lowest eigen values and their associated mode shapes. The critical load factor, α_{cr} is also determined and is the factor that the ultimate loading would have to be increased by to cause elastic instability. The factor α_{cr} is therefore the ratio of design loading to critical buckling load H_{crit} (European Committee for Standardization, 2005).

According to BS EN 1993 (2005), if the critical load factor α_{cr} is less than 10 for an elastic analysis, second order effects must be taken into account when calculating the ultimate capacity of the section.

A stability check must also be done to ensure that the design loading is less than the critical buckling load H_{crit}. The critical buckling load for an arch structure can be determined from Equation 4, where the coefficient C_1 is determined from Figure 16 and the length l refers to the length of the arch in plan (Parke & Hewson, 2008). The elastic modulus E and 2nd moment of inertia about the strong axis I_x are also required.

$$H_{crit} = C_1\,{EI_x}/{l^2}$$

(4)

Figure 16 is a plot for the critical buckling loads factor versus the rise to span ratio. The solid lines represent cross-sections of constant section properties and the dashes lines refer to cross-section of non-uniform cross section properties. Four arch articulation are considered in the graph, namely encastred (or hingless), one-hinged, two-hinged and three-hinged (Parke & Hewson, 2008).

Once the critical buckling load, H_{crit} has been determined, it can be compared to the ultimate axial load in the arch rib to evaluated whether he arch rib has enough strength and stiffness to resist buckling effects.

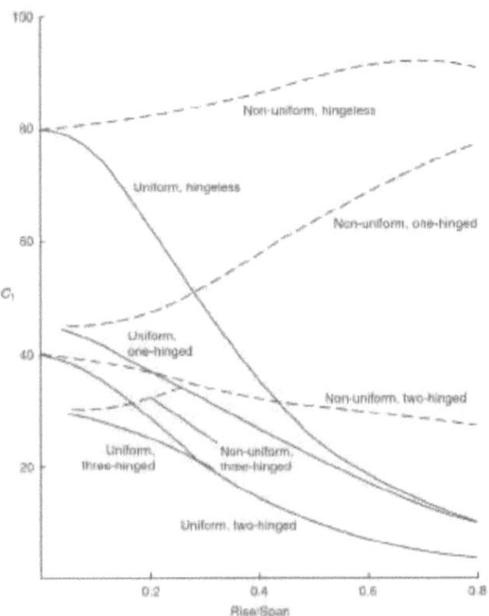

Figure 16 coefficients for in-plane buckling of parabolic arch (Parke & Hewson, 2008)

2.3.4. Prestressed Concrete Tension-Member

Concrete tension members are typically composed of a large concrete area, with bonded prestressed tendons and normally reinforced high tensile steel reinforcement. The large concrete area provides excellent axial stiffness, while the prestressing tendons provide the compression force in the member prior to the application of the external tension force. The normal reinforcement provides extra stiffness to the section in the event that cracking takes place.

Structural steel is often thought to be better suited for tension members, as it is known to have a better tensile capacity. Concrete on the other hand is known to be poor in tension and therefore it is thought of as an improper material for tension members. It is however found that prestressed concrete has been used as tension members in many applications, such as ties, truss member and hangers in bridges and soil anchors in retaining structures. Circular and parabolic applications are also found, especially for tanks, silos and pressure vessels. Some of the typical infrastructure applications are shown in Figure 17 (Naaman, 1982).

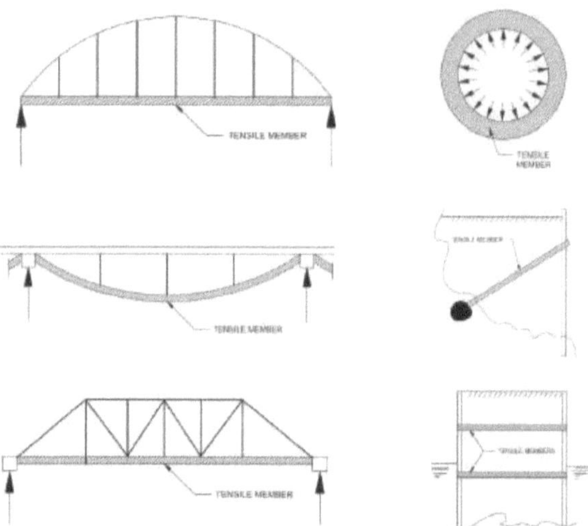

Figure 17 Typical applications of prestressed tension members (Naaman, 1982)

The reason for prestressed concrete's prevalence as tie-members is probably due to the high axial stiffness and inherent durability of concrete elements. It can be shown analytically that the axial stiffness, which is a function of the section area and modulus of elasticity, is significantly stiffer than structural steel for the same working load. This means that a prestressed element will experience much less elongation than a structural steel member designed for the same working load. This is especially beneficial for long span arch bridges, in which the sag of the arch is highly dependent on the distance between springing points.

Regarding durability, it is known that prestressed elements are always in compression and therefore shrinkage cracking can be limited under service conditions. The reduction of cracking means that the steel is protected from the ingress of detrimental particles/substances, which is important in members buried underground (Nielson, 1978).

When considering the behaviour of concrete tie-elements, it can be shown that the mechanics of a prestressed concrete tie-member working in direct tension is fairly simple in the linear elastic range of the material. However, the behaviour of the tension member is not as straight forward when the tensile strength of the concrete is exceeded and non-linear material behaviour of the composite section is experienced. This behaviour is demonstrated with a force-strain graph shown in Figure 18 and a more detailed example in Appendix A. It is therefore thought that most prestressed tie-members are designed so that they work in full prestress under service loads and is therefore in the linear elastic uncracked behaviour range.

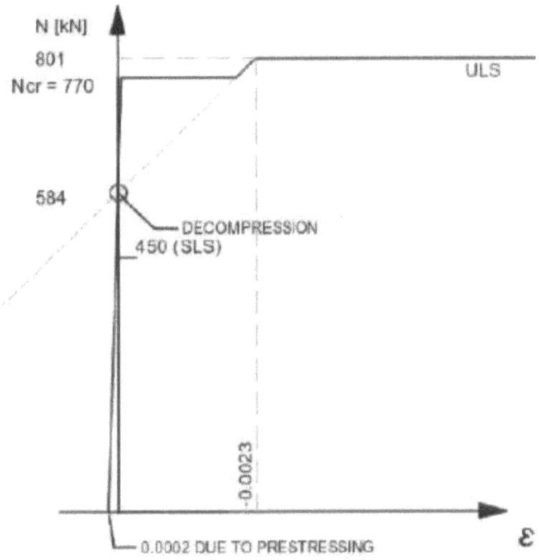

Figure 18 Theoretical stress strain behaviour of Tie-Member

It is however important to understand the non-linear behaviour of prestressed tie-members if there is a possibility that this partially prestressed state may be experienced. Wheen (Wheen, 1979) reported that while performing tensile testing of prestressed samples, there is a dramatic loud noise which occurs when the first crack of a prestressed tension member is experienced. This is due to the transfer of stress from the composite concrete section to the remaining prestressing steel and ordinary steel. During this event the axial stiffness of the section dramatically decreases, with the same external tension load, which results in a step in the elongation of the member. It is also interesting to note from Lin (1970), that generally

22

prestressed tension members have a very low strength threshold after decompression of the element has been reached. It is intuitive to assume that after decompression, the concrete still retains its own *low* tensile strength before failure of the concrete. Lin (1970) however makes the important observation that this assumption is only valid if the member has been constructed in a single monolithic cast and if there has been no restraining cracking through the depth of the member. If there are any such defects in the concrete, then the transfer of load from the concrete composite to only the steel will take place immediately at decompression and excessive elongations may be observed.

Finally, Wheen noted that tie members are well suited for structures wherein there is no substantial transient load variation, as large load variations in turn produce large tie-member extensions (Wheen, 1979). This is therefore favourable for a heavy, long span concrete tied-arch bridge, where the dead load equates to approximately 83% of the largest in-service load which is expected for the structure.

2.3.5. Classification of prestress members

Prestress concrete members may be classified according to the level of stress which the element will experience during its service life. The British design code BS 5400 Part 4 and South African Bridge design code TMH7 Part 3 have three different classifications for design:

- Class 1 – Full Prestress – No tensile stress permitted
- Class 2 – Limited Prestress – Tensile stress permitted, but is limited to the extent that no visible cracks develop
- Class 3 Partial Prestress – Tensile stresses permitted, but with surface crack widths limited to values prescribed by the particular code being used

Generally, the most critical limit state for Class 1 and Class 2 elements is the limit state of cracking, while the most critical limit state for Class 3 prestressed elements are often determined by the ultimate limit state or the limit state of deflection.

Liebenberg (1984) stated that traditionally engineers would favour a more conservative design and therefore prefer that prestressed elements do not crack. Sections that are designed to always be in compression, such as Class 1 members, are however costly, because it would require larger sections and more prestressing steel. On the other hand, partially prestressed members, i.e. Class 3 members, allow for cracking that must be controlled by the magnitude of prestressing force and the provision of non-prestressed reinforcing (Wiessler, 1984). It is therefore usual to design prestressed concrete members that fit in between these two extremes. Class 2 members require that the section is in full compression for the majority of its service life, but may develop limited tension stresses, but no visible cracking, when subject to the maximum design load at the serviceability limit state, after all creep and shrinkage losses have taken place (Robberts & Marshall, 2000). The Ashton Arch Bridge has been designed to comply with Class 2 criteria.

2.3.6. Partially prestressed concrete

Prestressed concrete originated in the 19th century, but was first patented by Eugene Freyssinet early in the 21st century. Originally prestressed concrete was intended to limit excessive cracking and deflections by providing compression to ordinary reinforced concrete (Wiessler, 1984). This method became popular in Europe after the Second World War when steel production declined (Hewson, 2012).

Liebenberg (1984) reported that the concept of partial prestressing was first introduced by Dr. Von Emperger of Viena, who performed test to show that cracked reinforced concrete could be improved by adding additional externally bonded pre-tensioned wires. Liebenberg (1984) further explains that the idea was advanced by Abeles and others in the early 1940's with much controversy in the prestressing fraternities, as it was still then a widely held view that prestressed concrete should remain only in compression. This may have been due to the lack of understanding of the possibility of corrosion of high tensile steel in cracked concrete (Wiessler, 1984).

According to Wiessler (1984) it was fortunate that several other engineers, including Leonhardt, Bachann and Lin continued developing the theory of partial prestressing and made it a more usable technique.

The method of partial prestressing essentially ensures that a structure remains in a compressed state for most of its service life, which is different to conventional prestressed concrete in that it is allowed to crack under the maximum service load. The prestressing force and ordinary reinforcement steel are then designed to limit the size of these cracks so that the durability and deformation characteristics can be controlled. No permanent deformations occur, as the material is still elastic and the cracks close up after the transient load is removed. However, at the ultimate limit state the section is overloaded and non-linear plastic behaviour is encountered in the concrete and/or steel. This condition results in irreversible deformations (Wiessler, 1984).

It is important to note here that the maximum service load rarely occurs and it is probable that the structure will remain crack free for the majority of its design life. Wiessler (State of the Art, 1984) states that "many structures designed by Abeles for the British Rail on the principle of partial prestressing were built in the 1950's and early 1960's and their performance to date is quite satisfactory." These structures were investigated in the 1970's and found to be structurally sound with no evidence of distress. Hewson (2012) also mentions that there is a possibility of a reduced durability in harsh environment, but that this has not been "highlighted" to date where partial prestress has been used.

Partial prestressing is beneficial in that there is a reduced post-tensioning force and better utilisation of the non-prestressed steel, when compared to full prestressing. This relates to economy. There are also quality related improvements as cracking and deflections are better controlled relative to normally reinforced concrete. Upward creep effects, which may exist in fully prestressed elements and downward creep effects, in normally reinforced concrete, are improved (Hewson, 2012). Partially prestressed concrete also behaves more ductile at the ultimate limit state, which is beneficial when the structure must resist abnormal loads, impact loads, differential settlement or extreme temperature variations (Wiessler, 1984).

After coming to the realisation that cracking in well compacted prestressed concrete is not as detrimental as first envisaged, the partial prestress philosophy eventually started gaining acceptance and has been adopted by many design codes towards the later part of the 20th century. In South Africa, bridge designers make use of the TMH 7 design code, which follows very closely the British design code BS5400. The British code makes use of the so called Hypothetical Tensile Stress Design Method (British Standards Institution, 1990), while the South African design code makes use of the Incremental Stress Limitation Design Method (Committee of State Road Authorities, 1989).

The Hypothetical Tensile Stress Design Method allows for the design of prestressing force so that the maximum tensile concrete fibre stress does not exceed an upper limit. The BS5400 code recommends a maximum tensile stress of up to 6 or 7 MPa (Hewson, 2012). The underlying principle is that by limiting the maximum tensile fibre stress, the maximum crack width is limited as well. This "deemed to satisfy" method was developed by Bates and Beeby in the late 1950 after performing extensive laboratory testing (Liebenberg, 1984). The hypothetical tensile stress method is thought to be the easier method of the two, since the analytical calculation assumes an uncracked section.

The Incremental Stress Limitation Design Method functions on a similar principle to the hypothetical tensile stress method. Here the stress increment in the post-tensioning cables is limited, and therefore the resulting crack widths is also considered to be limited. The method prescribes that the stress increment in prestressed and non-prestressed steel near the tension fibres should be limited to 75MPa and 150MPa for a 0.1mm and 0.2 mm nominal crack widths respectively. The stress increment is defined as the difference in the stress experienced by the steel during decompression of the concrete and under the action of the maximum service load (Committee of State Road Authorities, 1989). This method requires that the analysis be performed with a cracked section and therefore requires more computing effort. It may be beneficial to perform the design with the hypothetical stress approach and then verify the analysis with the stress increment method thereafter.

Modern codes such as the EUROCODE and AASHTO Standard specifications do not incorporate the different classification systems present in BS 5400 and TMH7. Instead AASHTO simply prescribes a maximum allowable tension stress for concrete during construction and under the maximum service load. It allows a maximum tension of $0.50\sqrt{f_{ci'}}$ for normal environments and $0.25\sqrt{f_{ci'}}$ for severe exposure environments, where $f_{ci'}$ refers to the 28 day compressive cylinder strength (Hewson, 2012). This equates to a 2.8MPa and a 3.2 MPa tension capacity for 40 MPa and 50 MPa 28 day concrete cube strengths, which appears to be a parallel to the Class 2 conditions, which allows for tension, but no cracking. The Eurocode does not specify an allowable tensile stress for prestressed concrete, but rather specifies that the area around the prestressing ducts be kept in compression under frequent load conditions. It further specifies that this compression area should be at least 100 mm away from the ducts. A second requirement is that the maximum allowable tension cracks developing in the tension zone should be smaller than 0.2 mm. This design criterion may be thought of as a parallel to the partial prestress philosophy (Hewson, 2012).

2.4. Construction considerations

The construction of complex long span structures requires comprehensive erection engineering during design stage and continuous technical support during construction. The Ashton Arch bridge has many visible complex systems which must be considered during the erection engineering. These include the installation and tensioning of the hanger cables, construction of high rising arch ribs and eventually the transverse launching of the 113 metre structure into its final position. The post-tensioning system is not visible to the eye, as it is hidden in the concrete deck, but it is equally technically challenging. The layout of the post-tensioning tendons is not conventional, since it comprises a grid of longitudinal and transverse tendons covering the 113 x 23 m deck area. To be exact, there are 64 transverse, draped tendons and 18 longitudinal, concentric tendons. All tendons were installed after casting the concrete segments making up the deck. This book is only concerned with 12 of the longitudinal tendons, which belong to the tie-beam elements. These 113 m long tendons required careful attention during installation and the procedure will therefore be discussed along with possible defective construction possibilities. It is normal practice to verify tendon elongations, but *lift-off* tests may be necessary if there are some doubts of the adequacy of the tendon elongation measurements (Corven & Moreton, 2013). Both the validation procedures will be reviewed.

2.4.1. Prestressed concrete

Prestressed concrete may be regarded as structural concrete in which internal compressive stresses have been introduced during construction to reduce potential tensile stresses which may arise from external loading (Robberts & Marshall, 2000). These internal compressive stresses may be applied by pre-tensioning or post-tensioning. Pre-tensioning refers to tensioning of individual strands prior to casting concrete around the strand. The strands remain in tension until the concrete has hardened sufficiently and are then released. Alternatively, the concrete may be cast with voided ducts. When the concrete has hardened sufficiently strands may be threaded through and tensioned. The later system is called post-tensioning and is the method of prestressing used on the Ashton Arch Bridge.

2.4.2. Installation of tendons

In early post-tensioned structures it is noted that prestressing tendons were pre-fitted into ducts at the manufacturer's warehouse and delivered to site as a preassembled unit. Strands were neatly arranged with spacers in between them prior to pulling the duct over the orderly arranged strands. At the construction site the heavy preassembled component was placed on rigid, closely supported stools inside the deck before casting the concrete. This heavy construction sagged notably between supports, resulting in significant "wobble" effect. Leonhardt (1964) notes that "it is becoming increasingly more general practice" to install void forming ducts in the concrete deck of long span bridges prior to installing the tendons. Leonhardt (1964) also notes that tendons were gripped with a device, such as a threaded sock and pulled through the duct with a mechanical winch (Figure 19).

Figure 19 a) Steel wire sock and b) mechanical winch for installation of strands

In modern times it is common practice for conventional small span prestressed concrete bridges to fix the duct in its final position and then draw strands through the ducts prior to casting the deck. This can easily be done one at a time by hand or alternatively it may be drawn with a make-shift pulling machine.

For larger bridge decks that are cast in sections, but are required to be stressed as a unit, it is not possible to install the full length of the tendon after the first cast. It is often then necessary to cast in the voided duct and install tendons at a later stage, when the full length of the deck is complete. In these cases, the tendons are either drawn through the duct void in a group of strands or they may be shot through strand by strand by mechanically driven rolls. While using either of these two methods, care must be taken to ensure that dirt does not enter the ducts with the tendons as this will result in additional unaccounted for friction (South African National Roads Authority Limited, 2011).

Long (1971) comments that it is known that tendons, made with strands in an orderly arrangement, have a natural lay, resulting in the outer strands rotating about 180 degrees over a length of 10 – 12 metres. This effectively means that a strand located at the bottom of the tendon at some point, will be located at the top of the tendon 10 - 12 metres thereafter. This rotation of the tendon inside the duct is not detrimental if the strands are installed in an orderly fashion. Long (1971) suggests that for tendons composed of many rings, it is beneficial to make the inner strands longer so as to assist with cable identification and positioning at the anchorages (Long, 1971).

Nawy (2008) adds further comments relating to the method of shooting strands one by one, also known as the *Push Through Method,* advising that it is beneficial to tension the individual strands, with a mono strand jack to a nominal stress. This is done to ensure that the strands are of the same length and uniform tension.

Shooting through strands pose additional risk to construction, but is often found to be more efficient for the contractor (Nawy, 2008). Cable shooting can damage the duct sheathing and therefore it is common practice to use a bullet nose on the tip of the strand projectile, which makes the tip blunt. For longer span structures and structures with many high and low points, shooting through strands may be problematic.

Therefore, for post-tensioned structures requiring installation of strands after casting the deck, the method of installation should be carefully considered. It is usually more economic to install strands by pushing and this is efficient for shorter spans. For longer spans it is necessary that the orderly arrangement of strands be guaranteed by careful consideration of the strand installation method so that the anticipated elongations can be achieved.

2.4.3. Tendon Elongation and friction losses

After the tendons have been successfully installed, the anchor head and wedges are placed on either side of the bridge deck and the tendons are ready to be tensioned. Prior to commencing post-tensioning, the site team should verify with the designer what the expected elongations are.

The theoretical tendon elongation can be estimated by utilising the simple linear material laws $\sigma = E_s \varepsilon$, where the stress σ is defined as F/A and the strain ε is defined as $\Delta l/L$. Combining these parameters and making Δl the subject of the equation, we find that $\Delta l = PL/AE$. This equation is valid for one dimensional object of constant force, area and elasticity along its length. It is however known that for post tensioned tendons, there are significant losses in force along the length of the cable due to the interaction of the strand with the duct (Robberts & Marshall, 2000). These are primarily friction losses and they must be allowed for when estimating the theoretical elongation of the tendon.

Friction loss

The magnitude of the tension force applied by the jack is not equal to the force experienced by the tendon during tensioning. This is because the jacking force is reduced by factors such as the elastic shortening of the concrete, friction in the jack, friction in the duct and anchorage seating (Robberts & Marshall, 2000). These instantaneous prestress losses are significant and must be accounted for in the analysis and design.

For the purpose of calculating tendon elongations, friction loss in the duct is the most significant prestress loss to consider. This loss of prestress force is due to friction between the tendon and the duct, which is constructed with an intended profile as indicated in Figure 20. The duct may however deviate from its intended profile during concrete placement or because the supporting stools are spaced too far apart, resulting in additional unintentional deviations. The magnitude of the friction force loss can therefore be broken down into friction losses resulting from intended angular deviation and unintentional angular deviations (wobble).

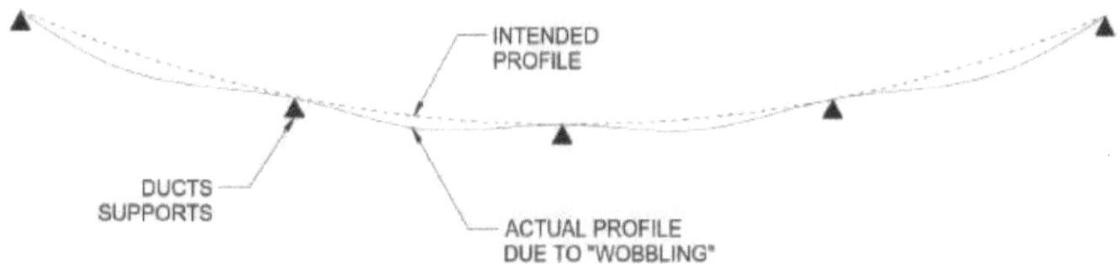

Figure 20 Wobble Effect (Robberts & Marshall, 2000, pp. 5-28)

Friction loss due to intentional angular deviation is graphically illustrated in Figure 21, where the tension force in the cable "P" induces a resultant normal force "$N = Pd\alpha$" in the concrete. The normal force between the concrete and tendon results in a frictional force tangential to the normal force of $f_1 = \mu\, Pd\alpha$. Additional frictional losses due to unintentional angular variations are accounted for empirically as a force per unit length $f_2 = KPds$. These two quantities are summed together and integrated over the length of the tendon so that the loss in force over the length can be determined. This varying tension force can be calculated using Coulomb's formulae (Robberts & Marshall, 2000) in Equation 5.

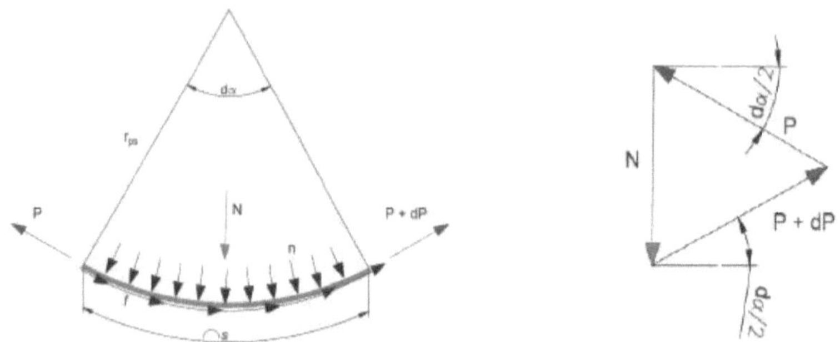

Figure 21 Friction loss due to intentional angular deviation

$$P(x) = P_0 . e^{-(\mu\alpha + kx)} \tag{5}$$

The friction coefficient μ has generally been found to be 0.3 for a strand in a steel duct, but may be obtained from the manufacturers' literature for the system. The sum of deviations α over the distance x is dependent on the intended angular tendon profile. The recommended wobble factor K for properly supported duct formers is 17×10^{-4} radians per unit length (Committee of State Road Authorities, 1989). The friction coefficient μ takes into consideration jamming forces exerted by internal strands onto the periphery strands, which are in contact with the steel duct. These additional forces arise by their arrangement in relation to each other and Leonhardt (1964) described it as a necessary evil for multi-strand tendons. Berverly (2013) similarly refers to a squeezing factor, which is a function of the degree of filling of the duct and increases the physical friction coefficient μ_0 by a factor of 1.3 - 1.35 for tendons filled with 50 - 55% steel area relative to duct area. This squeezing factor is however already allowed for in the design codes.

When plotting Equation 5 over the length of a tendon, a force profile similar to Figure 22 is obtained. The figure illustrates the gradual decrease of force in the tendon between the live anchor on the left-hand side and the dead anchor, on the right-hand side. The figure also illustrates loss in force due to anchorage seating after locking off the tendon when tensioning is complete.

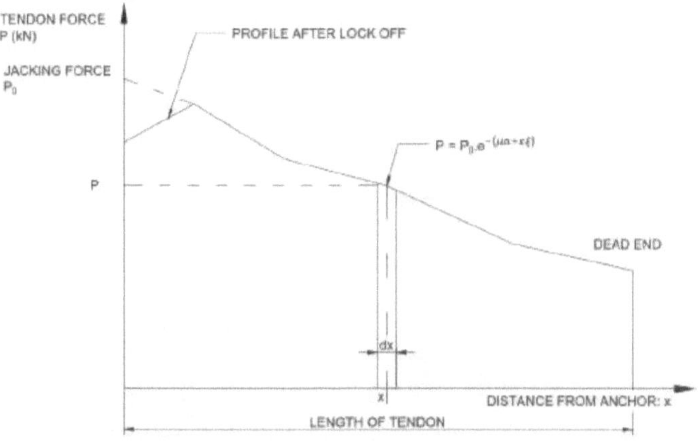

Figure 22 Tendon Force Profile (Hewson, 2012, p. 69)

After the force profile has been plotted, the theoretical extension may be calculated by integrating the force over the length of the duct and dividing it by with area and stiffness modulus of the tendon, as shown in Equation 6. This varying force may of course also be interpreted as the geometric area under the graph (South African National Roads Authority Limited, 2011).

$$\Delta l = \left. \int_0^L P(x).dx \middle/ (A_s.E_s) \right.$$

(6)

The elastic modulus E_s used in the calculation of the theoretical elongation is determined by carrying out tests on each batch from which the strand was drawn. It should however be noted that the tests conducted are for short lengths of strands, which are fairly stiff. For longer strand, the outer spiralling lay wire in the strand reduces in stiffness with longer lengths. It is therefore possible that the E_s may be as much as 5% lower for long lengths of multi-strand tendons than the value reported by the manufactures (Hewson, 2012, p. 76).

Measuring Physical Elongations

After the theoretical elongations has been verified the post-tensioning may commence. During the post-tensioning operation, the multi-strand tendon is restrained by wedges in the dead anchor on the far end of the bridge and wedges in the hydraulic jack at the near end. The hydraulic jack pulls on the tendon so that the tendon elongates away from the deck. Measurements are recorded from the jack piston as it moves away from the stationary body of the jack (Figure 23).

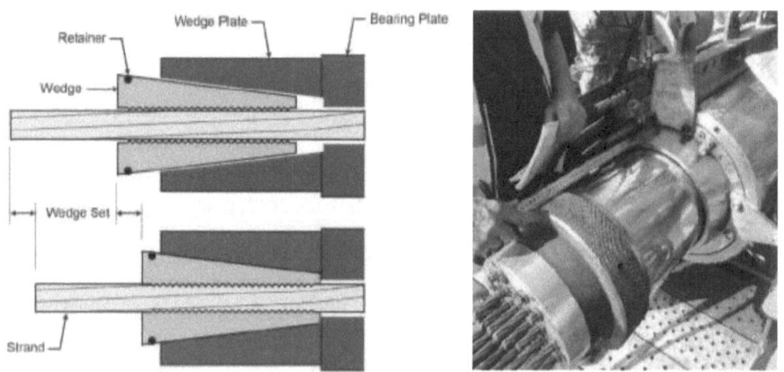

Figure 23 Wedge set (Corven & Moreton, 2013) and measurement of tendon elongation

The theoretical elongation calculated by Equation 6 may then be compared with the "face to face" physical elongation, which is the elongation of the tendon between the anchorages. The "face to face" elongations can be determined by subtracting the elongation of the tendon in the jack and the draw-in of the wedges at the dead anchor from the total elongations measure at the end of the jack. This principle is demonstrated with Figure 24.

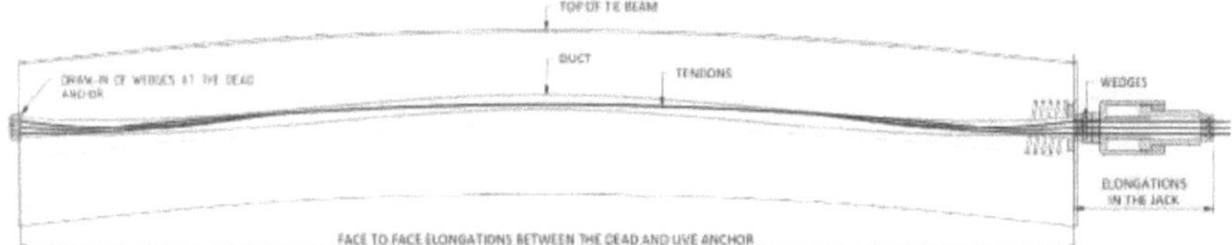

Figure 24 Calculation of "Face to Face" Elongations

The COLTO specifications used in South African road construction prescribes that the maximum difference between measured "face to face" elongations of individual tendons shall not exceed ± 6% of the theoretical extensions. Additionally, the average measured extensions of a group of tendons in a structural member should not exceed ± 3% of the theoretical extensions (Committee of Land Transport Officials, 1998).

It is however often found that the measured elongations do not match what was calculated theoretically and then it becomes necessary for the team to investigate the reason for these deviations. A deviation from the theoretically calculated elongation generally indicate an over or underestimation of the friction coefficients, wobble coefficients or elastic modulus of the strand. Significant deviations, greater than say ±8%, are however more problematic and may imply defects in the construction of the duct or strand. This is especially true if there are outliers in a group of tendons.

Two possibilities exist here, namely greater elongations than anticipated and lower extensions than anticipated. Significantly greater elongations or over-extensions are often due to slippage of wedges at the dead end of the tendon. This can however be verified by checking that the wedges in the dead-end anchor plate are properly seated.

Lower extensions or under-extensions usually indicate blockages in the duct which then prevents the full length of the tendon elongating. This is usually due to concrete entering the duct during casting of the deck, but may also be as a result of a knot formations or misaligned ducts at the male/female connection point (Corven & Moreton, 2013).

2.4.4. Uniformity of prestressing in the strands

Another defect which could occur if tendons are not installed correctly is the possible occurrence of non-uniform stresses among the individual strands due to the disorderly configuration of the tendons. This may arise by two different mechanisms:

- Different strand lengths (Leonhardt, 1964)

The initially installed strands may be substantially longer than the finally installed strands, which are straighter. A helix shape adds additional length relative to the centre line of the duct because of the natural lay of the strand. This effect becomes more pronounced as the length of the duct increases. It is therefore necessary to ensure that all the strands in the duct are of the same length so that the stresses may be uniformly distributed among the strands.

- Formation of a knot (Kreger, 1998)

It has been noted in case studies of long post-tensioning tendons that strand tangling located along the length of the tendon could be pulled into a knot near the anchorage as shown in Figure 25. A saddle point located close to the anchorage position could also prevent the strand from disbursing the twist over a greater distance and therefore could contribute to a knot formation and under-elongations. Such a knot could result in a loud clicking noise during tensioning as the tendon attempts to untangle itself and can cause an uneven distribution of force among the strands of the tendon.

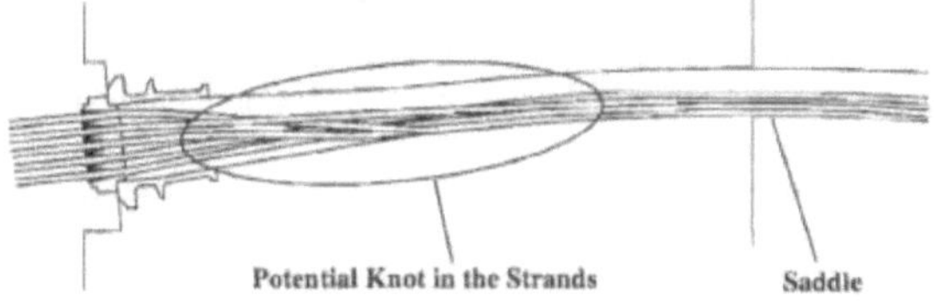

Figure 25 Knots may have occurred at the saddle positions (Kreger, 1998)

Tendons with uneven prestress distribution could become overstressed and break as the final force in the tendon is being applied. Cognisance should therefore be taken to ensure that all strands are of the same length and free of entanglement.

2.4.5. Lift-Off Testing

Lift off testing has been used in a variety of applications to verify the force in tensioned cables. Typical examples include rock anchors, walls of containment buildings such as nuclear power plants and bridge decks. The test can however only be performed on tendons before grouting, when the tendon is taut, but free to move inside the duct. It is also normally preferred on straight tendons where there is minimal friction loss in the cable.

The test involves setting-up the hydraulic jack on the anchor head of an already tensioned tendon, as if to tension for another tensioning stage. The jack wedges engage with the tendon strands and the jack pressure is gradually increased while closely monitoring the gauge pressure and elongations. In a Lift-Off Test the important aspect to capture is the exact pressure when the block disengages from the anchor head. To achieve this the author makes distinction between the two clearly defined stages, namely "before lift-off" and "after lift-off". During these two stages, a linear trend is observed in the force-extension diagram shown in Figure 26 and Figure 27. While the force is low, it is assumed that there is still contact between the anchor head and the bearing plate. At a higher pressure, i.e. "after lift-off" it is assumed that the anchor head has lifted off the bearing plate and there is a "gap" between the two elements.

When examining the force-extension diagram it is noted that there is no clearly defined point of lift-off, but rather a gradual lift-off of the anchor head illustrated by a curved line between the two linear stages. The lift-off force is therefore a theoretically determined value and is defined as the force at the intersection of the two linear lines. (Nuclear Energy Agency, 1997)

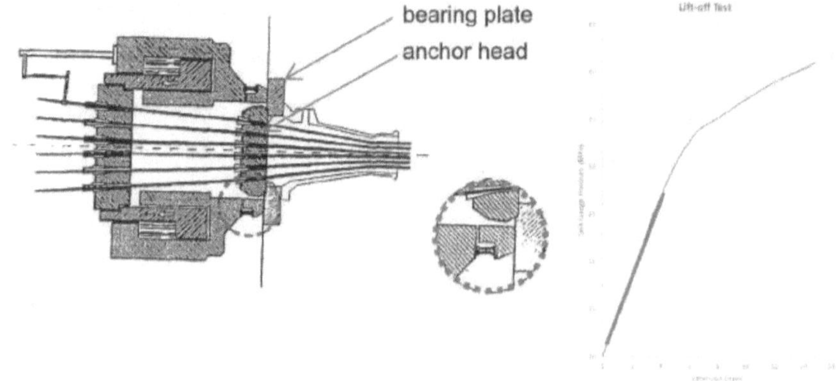

Figure 26 Before Lift Off Testing

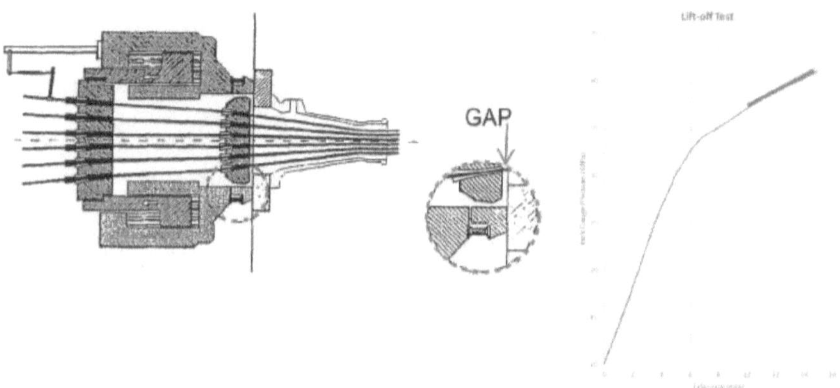

Figure 27 After Lift Off Testing

An example of where lift off-testing was performed is shown in Figure 28 on re-stressable rock anchors in the Kaaimans Pass Viaduct. Five strands were fixed in the anchor head by means of wedges. The jack is placed on a spacer stool so that it does not interfere with the anchor head.

The lift-off extensions are measured by means of a dial gauge mounted on a sturdy tripod behind the jack. The 12 metre unbonded tendon has a straight profile and therefore the slope of the pressure-extension graph has a constant slope after lift-off has been reached. (Kruger, Newmark, & Smuts, 2008).

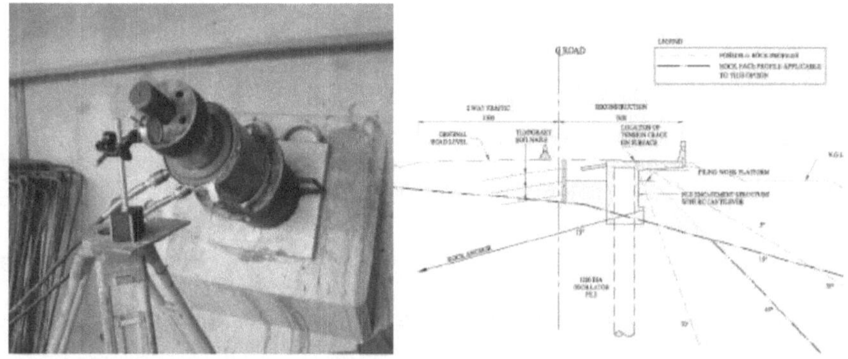

Figure 28 – Lift-off test on Kaaimans pass (Kruger, Newmark, & Smuts, 2008)

Vives and Ubalde (1997) reported on anchor lift off testing of a 37 T 15 anchorage type by Freyssinet on the Asco Nuclear Power Plant as part of routine maintenance of a containment building. The un-grouted tendons are located on vertical walls and a spherical dome. Measuring of the prestressing force was done with a 1000-ton capacity jack. Strands were inserted through the jack and thereafter the jack wedges lock-in on the ends of the strands. The jack is placed up against a bearing plate and the force is gradually increased until the anchor block lifts off from the bearing plate as illustrated with Figure 26 and Figure 27.

3. ASHTON ARCH BRIDGE

3.1. Bridge description

At the time of writing, the New Ashton Arch Bridge was being constructed in the town of Ashton, South Africa on Route 62 crossing the Cogmanskloof river. The bridge replaces an existing solid spandrel multi-arch bridge which was built in the 1930's and retrofitted in the 1950's, but has now reached the end of its functional service life. The author is a senior structural engineer on the project, responsible for the calibration of the FEM model, assistance with erection engineering and providing technical support to construction teams.

The new bridge is a tied arch structure with a cable-supported concrete deck which spans 110 metres between supports. The arch rise above deck level is approximately 23 metres. The deck is 24 metres wide and comprises a grid of post tensioned concrete beams, topped with a 300 mm thick reinforced concrete deck slab. A three-dimensional rendering of the structure in its complete state is shown in Figure 29.

The horizontal thrust force of the reinforced concrete arches is retained by post tensioned concrete tie-beams that are integral with the bridge deck. The hangers are inclined and comprise locked coil cables with fork sockets at either ends, which are connected to the tie-beams and arches by means of special cast steel connection anchor. The lateral stability of the arch ribs is controlled by five stabiliser beams, four of which are placed along the sloping parts of the arches and one at the crown.

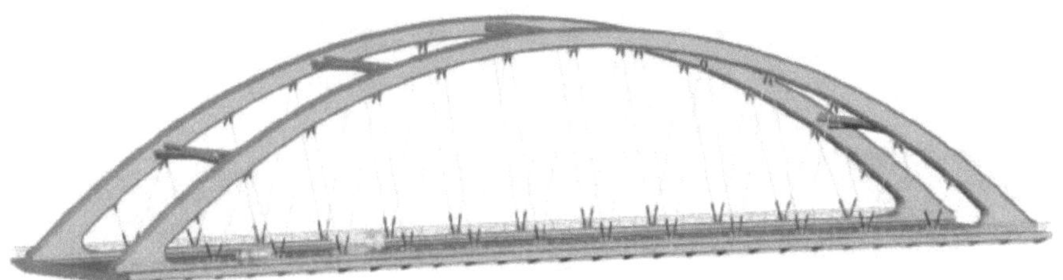

Figure 29 Architectural impression of Ashton Arch Bridge

The new bridge will be constructed downstream of an existing bridge, in its temporary position on temporary falsework. Once complete, traffic will be diverted onto the new bridge in its temporary position, to provide full access for the demolition of the existing bridge and construction of the new permanent abutments. In the final construction stage, the bridge will be launched transversely into its final position over the space of a weekend.

3.2. Design considerations

The new Ashton bridge has a span to rise ratio of 5:1 and a total dead load of approximately 77 MN. This is large dead load and is shared evenly among 4 bearings, which carry approximately 19.3 MN each. Using the preliminary design calculation an equivalent uniformly distribute load q can be calculated using Equation 1 and the resulting horizontal thrust H can be calculated using Equation 2. The thrust calculated in the tie-beam will be relatively equal to the thrust in the arch.

$$q = {}^{2V}\!/_L = {}^{2\,x\,19\,260}\!/_{110} = 350,182\ kN/m$$

$$H = {}^{qL^2}\!/_{8f} = {}^{(350.182)(110^2)}\!/_{8\,x\,23} = 23\,028\ kN$$

When comparing this value to the detailed FEM analysis we find that the thrust in the tie-beam are very similar, with the FEM analysis indicating a force in the Tie-beam varying between 24 000 MN and 21 300 MN. The varying magnitude tie-beam thrust is primarily because of post-tensioning friction losses. The values quoted refer to the structural behaviour of the members at the end of construction and these values will reduce by about 10 – 15 % due to creep and shrinkage throughout the design life of the structure.

3.2.1. Compression and tension members

The arch ribs were sized for the ultimate limit state (ULS) by considering axial loading and biaxial bending using TMH 7 Part 3 (Committee of State Road Authorities, 1989). Since the critical load factor was found to be less than 10, second order effects were taken into account in the design. The lowest eigenvalue and mode shape were extracted from the FEM eigenvalue analysis and is shown in Figure 30. This mode shape was used to derive effective lengths for the ULS capacity analysis. The ULS capacity of the tie-beams were also designed for using envelopes of ULS combination according to TMH7 Part 2. The resulting cross sections are shown in Figure 31.

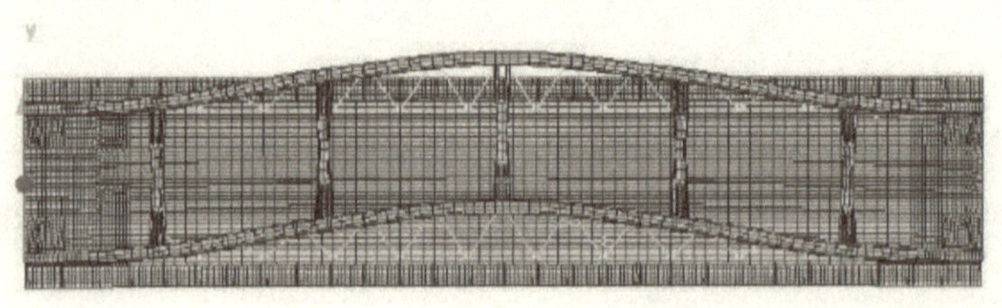

Figure 30 Lateral arch bending - 1st mode shape

All prestressed concrete members were also checked to ensure that the fibre stresses experienced by the structure were within Class 2 limits under the Serviceability Limit States (SLS) conditions. The method of construction, where the deck is discontinuous during prestressing (Figure 43) plays a significant part in the final built-in stresses at completion of construction. An envelope for maximum and minimum stresses were extracted for extreme fibres for the stages *during construction* and for the *in-service* period, where long term creep and shrinkage is summed with a SLS load combination envelope.

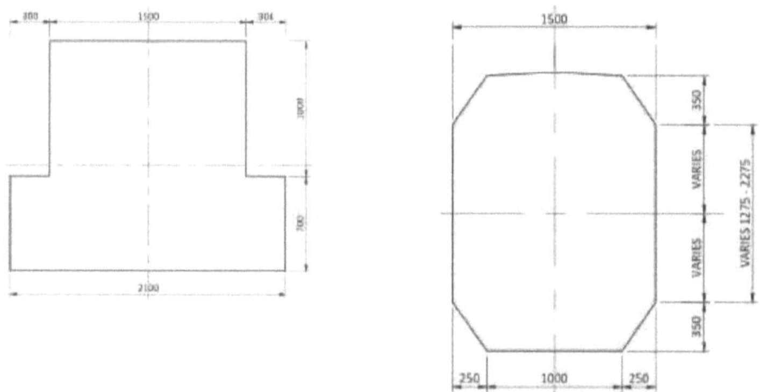

Figure 31 Cross section of Tie-Beam (left) and Arch Rib (right)

3.2.2. Bearing articulation

Globally, tied-arch bridges behaves similarly to simply supported bridges in terms of its interaction with supports. The permanent bearing articulation of Ashton bridge is presented in Figure 32. Free movement is allowed in the longitudinal directions, which allows for long-term creep and shrinkage and thermal movement of the deck. The deck is fixed at one end so that horizontal transient loads, such as wind, breaking, skidding and flood loads can be restrained in the lateral direction.

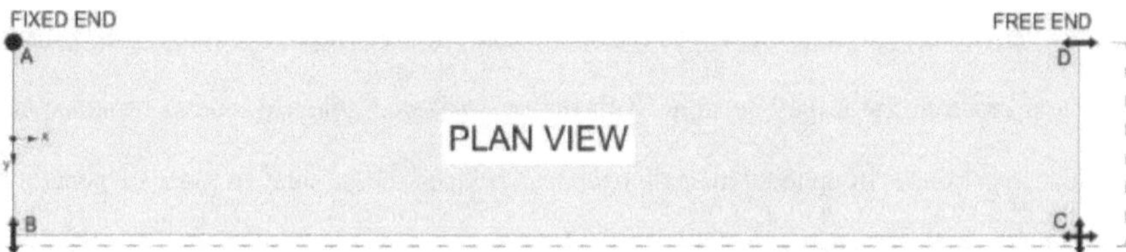

Figure 32 Articulation of typical tied-arch bridge

The bearings may be modelled in a FEM software package with a rigid link and an elastic link. The rigid link changes the point about which the tie-beam flexure takes place. The elastic link models the specific bearing and allows a predefined stiffness in the lateral restrained direction. The bearing modelled for the Ashton Arch Bridge are shown in Figure 33.

BEARING STIFFNESS MODELING ASSUMPTIONS (kN/m)				
Positions	Type	Kx	Ky	Kz
A	Fix - pot	1E5	1E5	1E7
B	Uni-directional	1E5	0	1E7
C	Uni-directional	0	1E5	1E7
D	Multi-directional	0	0	1E7

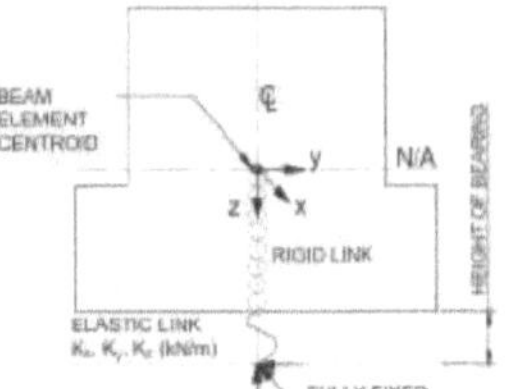

Figure 33 Bearing modelling

3.2.3. Creep and Shrinkage

The arch ribs and tie-beams members transfer large normal forces during their design life. These members are therefore susceptible to long-term creep and shrinkage (C&S). The arch rib will tend to compress and sag due to creep shortening of the member over time. The tie-beam, which is directly connected to the arch, via the hangers will also sag as the arch crown drops. Therefore it is important that the tie-beam has an adequate pre-camber to ensure that it never drops below the horizontal during its design life, as this would not be aesthetically pleasing.

Rilem Report TC 107-GCS (Bazant & Baweja, 1996) states that laborious creep and shrinkage analysis is not required for all types of structures. The document defines 5 levels of creep sensitive structures, based primarily on the span length of the bridge. The report states that a structural creep analysis may be done with the *age-adjusted effective modulus method* for medium span bridges of up to 80 m and longer span bridges. These analysis tools are generally available on bridge analysis software packages, such as Sofistik, RM Bridge and Midas. The age-adjusted method used in these software packages is demonstrated in Figure 34 and Figure 35. During a construction stage analysis, each element which is active at the time of the *Creep-Step* evaluation is analysed for creep and shrinkage effects, for the duration of the C&S period.

The software makes use of a C&S prediction model (Figure 35) to estimate the strain generated for that period, using the age of the element T_1. The age of the tie-beam at each analysis stage is shown in Table 2. In Figure 34 the various construction activities are plotted on a time scale. The tie-beam and deck slab are cast at time T_0 as indicated. The elements stiffness's are activated at the at CS10 with their self-weight load. An additional load case, CS11 is created for application of post-tensioning load. Thereafter the load case CS15 is create for C&S for the time period between CS20 and CS10/CS11. The resulting stress state of each load case is summed by the theory of superposition in a cumulative manner.

Table 2 Typical stage planning for C&S model input

CS No	Construction stage description	$T_1 = T_0+T$ (days)	Activation	Loading	C&S Period
10	Activate self-weight of tie-beam and deck slab		Self-weight		
11	Post-tensioning			Post-tensioning	
15	*Creep-Step* until CS20	21			14 days
20	Activate self-weight of arch ribs & cantilevers		Self-weight		
25	*Creep-Step* until CS30	35			14 days
30	Tension Hangers			Hanger tensioning	
35	*Creep-Step* until 30 years	10985			10950 days

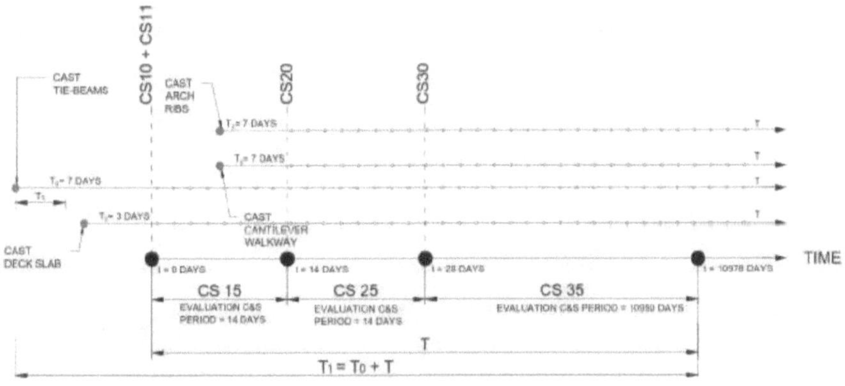

Figure 34 Age adjusted method for C&S

46

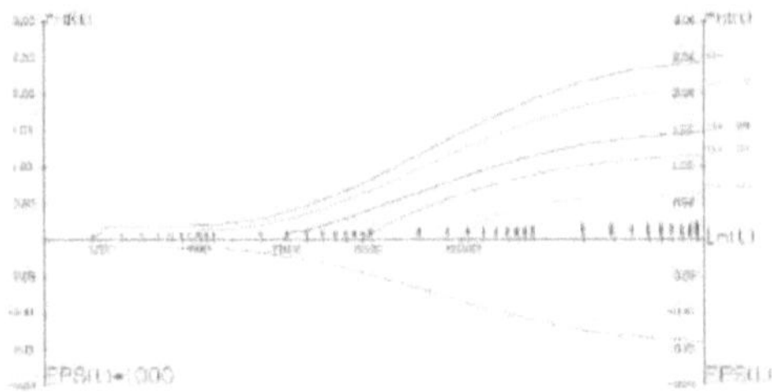

Figure 35 Creep and shrinkage strain calculation by Bentley RM Bridges

3.3. Construction considerations

3.3.1. Deck construction sequence

The deck of the arch bridge is built on temporary staging in the river bed, adjacent to its final intended position. The staging below the concrete consists of two layers of smooth shutter-board to reduce friction and encourage longitudinal sliding during post tensioning. The deck is constructed in eleven pours, including the two tie-beams with their associated massive spring points. These tie-beam members are cast during Pours 1, 4 and 8 and Pours 2, 5 and 9 respectively. In-between the tie-beams are three parallel smaller longitudinal beams and transverse beams, covered with a 300 mm thick slab. It is intended that the deck slab be left incomplete at the time of tie-beam prestressing, so that the stress generated at the head of the beams flows predominantly into the tie-beams. This is the so called "hit and miss" configuration and is illustrated in Figure 36.

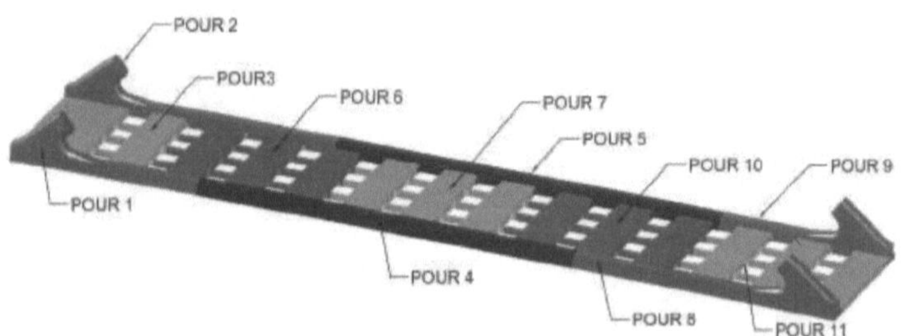

Figure 36 Deck construction sequence

After successful prestressing of the tie-beams, the infill deck slab will be cast and there after post-tensioning of the intermediate longitudinal beams and transverse beams will be done. The delayed longitudinal prestressing will add compression into the deck slab, so that the reverse thrust action of the arch does not cause excessive tension in these infill slabs.

3.3.2. Installation of tendons

The tie-beam members each consists of 6 longitudinal tendons, where each tendon comprises of 31 strands that have primarily straight profiles as shown in Figure 37 and Figure 38. Voided ducts were cast into the tie beams during the construction of the deck. The internal diameter of the duct is 120 mm, which is recommended for a 31-strand tendon constructed on site (European Organisation for Technical Approvals, 2013). It is noted that the internal diameter for a tendon which is pre-manufactured is 115 mm, which implies that the size of the duct is a function of the method of installation. If the tendons are post-installed correctly, without entanglement, the squeeze factor should not affect the friction losses significantly.

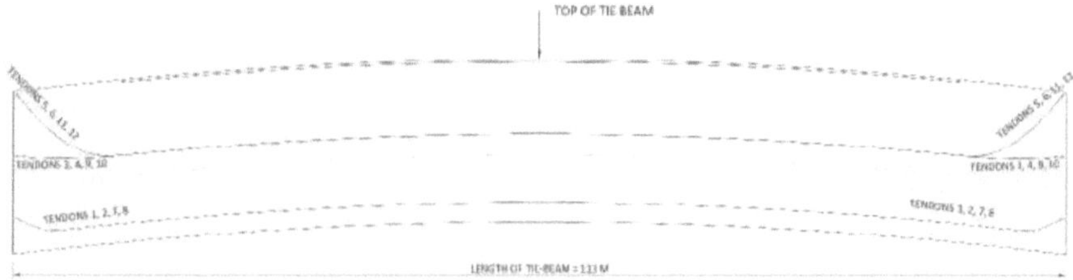

Figure 37 Tie-Beam Profile on long section with distorted length to height scale

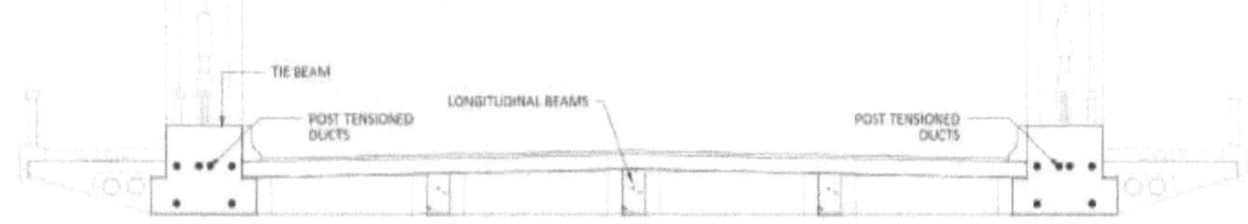

Figure 38 Cross-section of deck indicating position of tendons in the Tie-Beam

The push-through method was used to install the tendons, which means that individual strands were pushed through one at a time. During the installation of the strands by this method it was noted that the strands, particularly the first few, form a helix around the inside perimeter of the steel duct as it propagates through the length of the duct. This phenomenon is experienced when uncoiling the strand directly from the bundled coil, which results in the inserted strand following the lay of the bundled coil.

Site staff reported that it was very difficult to install the final few strands, of the 31-strand tendons, and therefore the pull-through method was also employed for these strands. It was reported previously that excessive force required to install a strand is indicative of unwanted entanglement (Breen, Davis, Frank, & Tran, 1993).

3.3.3. Tensioning procedure

The tied-arch bridge is 113 metres long and must be post tensioned from both sides of the deck because of excessive friction losses anticipated. The post-tensioning activity for a single tendon can be broken down into four stages:

- Tension from Jack Point 1

- Seat Wedges at Jack Point 1

- Tension from Jack Point 2

- Seat Wedges at Jack Point 2

Calculation of the force losses for such a complicated profile and stressing sequence is made easier by using computer software, such as Sofistik and the resulting elongations are then computed with the same principles described by Equation 5 and Equation 6. The anticipated force in the tie-beam tendons has been plotted for tensioning from Jack Point 1 and 2 in Figure 39, where the magnitude of the force in the tendon is on a normalised scale.

During tensioning from Jack Point 1, the force in the tendon is taken to its maximum at the position of the jack, known as the live end. There is a loss in force along the length of the tendon due to frictional losses discussed in Section 2.4.3. After reaching the desired tensioning force, the cables are locked-off and the wedges are allowed to seat. When the wedges seat there is a notable drop of force in the tendon at the live end.

Tensioning then carries on at the opposite end of the deck where the tendon is already highly tensioned. The force in the jack is increased until it reaches the force in tendon and thereafter the wedges disengage. The pressure in the jack is then further increased until it reaches the design pressure and then the tendon is locked off as described with the first pull.

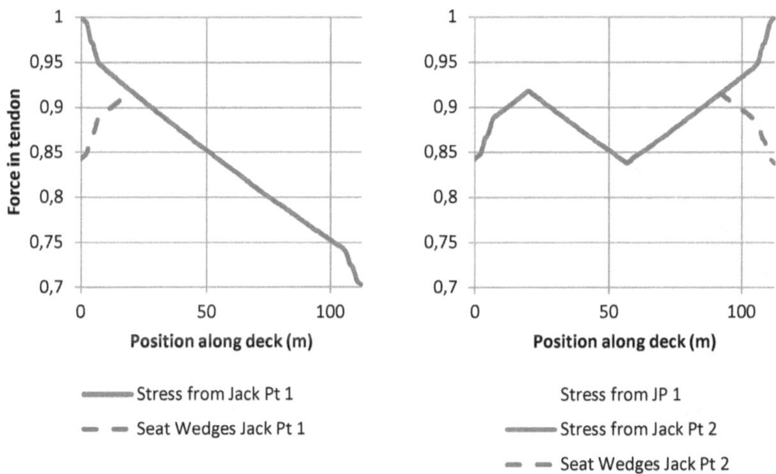

Figure 39 Force in Tendon after tensioning from JP1 and JP 2

3.3.4. Problems encountered during prestressing

The tie-beam tendons have a predominant straight profile with the cumulative intentional angular deviation of approximately 0.3 radians. The tendon has a straight profile with the main angular deviation at the ends to accommodate the anchor heads. With a fairly low angular deviation in comparison to the usual draped continuous girder applications, it was anticipated that the design elongations would be achieved with no trouble. This was however not the case.

Two tendons were initially tensioned and significantly lower extensions were encountered than anticipated, namely 556 mm instead of 692 mm and 588 mm instead of 714 mm. This is a 20 % and 18 % under extension, which significantly exceeds the 6 % tolerance specified by COLTO (Committee of Land Transport Officials, 1998). A typical under-elongation is illustrated in Figure 40. The post-tensioning activity was halted and investigations were undertaken to establish the cause of the under-extensions. These investigations included tensioning trials of the remaining tendons without the wedges in the anchor head, so that the wedges could not lock-off after the test.

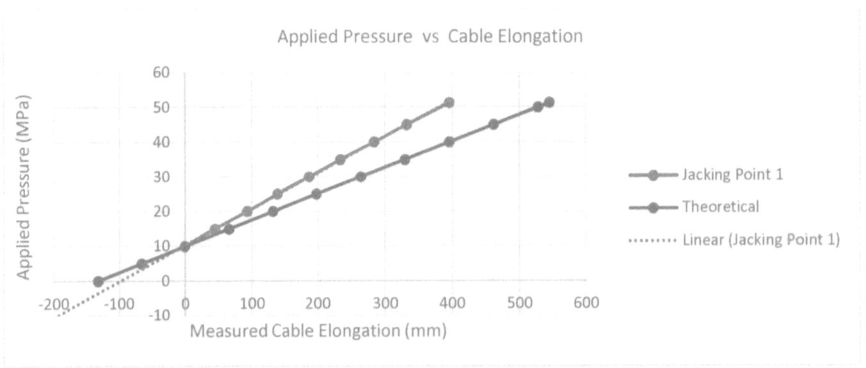

Figure 40 Elongation comparison

The test tensioning revealed that all the installed tendons appeared to be consistently stiffer than anticipated, which implies higher friction losses or a blockage in all the ducts.

A sensitivity analysis was carried out using Equation 5 and Equation 6 to determine the theoretical elongations for various friction and wobble coefficients. The coefficients that were required to correlate theoretical elongations with actual elongations were however unrealistically high and were therefore discarded as the root cause of the under elongations. It was therefore speculated that the method of installation had resulted in the disorderly arrangement of the strands and had thus created a blockage by the entanglement of the same strands.

Lift-off tests were then performed at the dead-end of the two already tensioned tendons to determine the transmission force. It was found that the force transferred was significantly lower than the design and therefore confirmed the excessive friction loss which was initially flagged by the low elongations. The information obtained by the lift-off test was also used to calculate actual friction coefficients, but these values were later deemed to be irrelevant because of the defective entangled strand. This is because Equation 5 assumes an orderly arrangement of the strands inside the duct, such that the length of the strand is approximately perpendicular to the corrugated folds and therefore the contact area between the strand and the corrugated steel duct is a minimum.

It was therefore necessary to de-tension the tendons already tensioned and remove them from the ducts. Strands were then reinstalled with a method that would ensure that the different strands are orientated in an orderly manner. This was done using the Pull Through Method where strands were pulled in bundles of 7 strands with a mechanical winch.

Strands were pulled with a winch and steel wire sock as shown in Figure 19. For a duct with 31 strands, the winch pulled 7 bundles to complete the tendon as indicated in Figure 41.

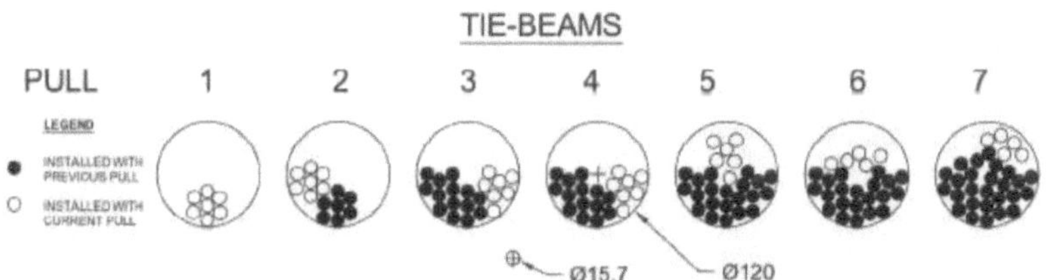

Figure 41 Pull Through Installation sequence

4. TIE-BEAM BEHAVIOUR VALIDATION METHODOLOGY

Verification may be defined as the process of determining that a model accurately represents the designer's conceptual description and its solution. *Validation* is defined as the process of determining the degree to which the model represents the real world. These processes allow models to make engineering predictions with quantified confidence (Zong, Lin, & Niu, 2015).

The structural behaviour parameters that must be validated in this research book are the *Input* and *Output* parameters as defined in Figure 42. This includes the post-tensioning forces applied to the tie-beam and the structural response of the tie-beam. These validations are done in three steps:

- Calculation of the theoretical behaviour

- Measurement of the actual behaviour

- Comparison of the data to validate the model

4.1. System behaviour validation

At this stage it is useful to define what is meant by structural behaviour.

The structural behaviour of a static system, subjected to external load, can be described by the forces and displacements that arise at all points in the system due to the external load. For a given load, these structural responses are dependent on the structural system stiffness (i.e. boundary conditions, geometry and material properties). This is graphically illustrated with an input-output diagram in Figure 42.

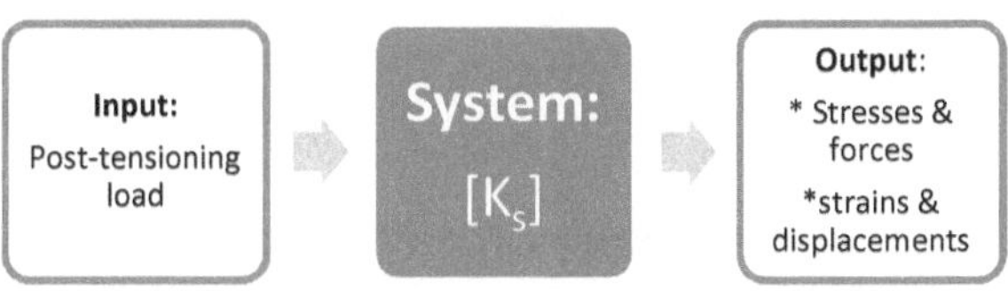

Figure 42 Structural Behaviour

The structural behaviour was monitored during the post-tensioning of the tie-beams, which was the 1st major construction event identified. The *input* and *output* structural behaviour parameters required validating by comparing theoretical results with physical measurements recorded during this construction event. The detail of each parameter evaluation is described below.

4.1.1. Input:

The *Input* load to the system is the post-tensioning force applied to the tie-beams by twelve 112.5 m long tendons. For validating the structural behaviour, it is important that the correct load is applied to the structure.

The theoretical model calculates the friction losses along the length of the tendon using Equation 5 and therefore the tendon force along its length is known. Field measuring must be performed to ensure that the design force of the tendons is realised by the structure. This can be achieved by monitoring the force applied by the hydraulic jack during the tensioning of the strands. It is equally important that the tendon elongations measured during tensioning are compared to theoretical elongations, as this is an indirect assessment of the accuracy of construction of the ducts and the tendons. These checks will ensure that the force transmitted from the tendon to the concrete deck is as anticipated by the designer. If the measured elongations are vary significantly from what was anticipated, lift-off testing may also be performed.

4.1.2. System:

The *System* relates to the geometrical arrangement, boundary conditions and material properties of the bridge during the post-tensioning (PT) event. In the theoretical FEM model this can be mathematically expressed with the stiffness matrix which is assembled from the finite element members.

The geometric configuration of the bridge at the time of post-tensioning is complex and has been designed to allow maximum post-tensioning force to be attracted to the tie beam during tensioning. A plan view of the deck with nomenclature is shown in Figure 43.

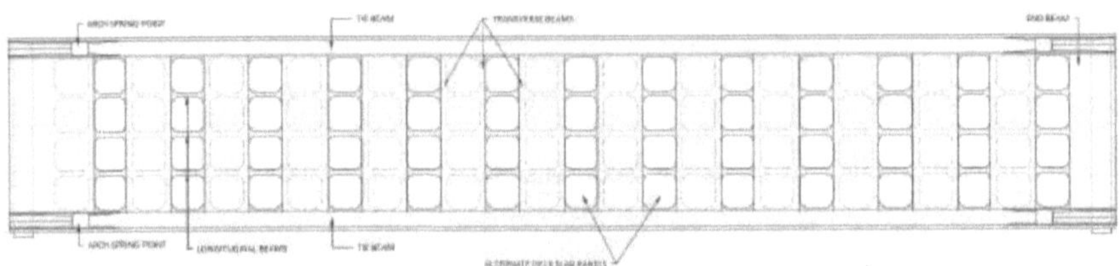

Figure 43 Plan view of deck configuration during post-tensioning of Tie-Beams

During FEM *model updating* the geometric properties of the finite elements, elastic modulus, as well as the bearing stiffness assumptions are updated to reconcile differences in the model prediction (Zong, Lin, & Niu, 2015). The geometric arrangement is however regularly checked by routine quality control procedures during construction of the bridge. Additionally, the material properties were tested at the University of Cape Town's laboratory and these results were input into the designer's FEM model. The FEM model is therefore regarded as accurate for these parameters.

Only the boundary conditions may require updating to reconcile differences encountered when validating the input and output parameters. This *model updating* procedure is however beyond the scope of this book.

4.1.3. Output:

The response of the system is usually measured in terms of strain and displacements. Once the force along the tendon has been calculated, the tendon force is transformed and applied to the centroid of the concrete beam segments as an equivalent joint force (Scordelis, 1984). This joint force can be transformed and applied to the beam as a varying distributed load in the FEM model. The FEM model will calculate the theoretical strains and displacements with conventional finite element theory.

Field tests can be used to validate these theoretical calculations. This includes measuring the geometric deformation with a surveying equipment and strain measurement at vulnerable zones.

4.2. Load Validation

4.2.1. Tendon elongations measurement

Calculation of the theoretical tendon elongation

The design of the tie-beam requires that the 6 tendons be tensioned to 75.0 % of the ultimate yield strength (F_{ptk}). Each tendon consists of 31 strands with a cross-section area of 150 mm^2 per strand, which when stressed to 75.0 % of their ultimate yield strength will produce a tension force of 6486 kN. The initial jacking force P_0 may then be calculated as follows:

P_0 = (no. strands x A_{steel} x F_{ptk}) 75.0 % = (31·150·1860)75.0 % = 6486 kN

The elongation is then calculated with Equation 5 and Equation 6, assuming the design friction coefficients and using the elastic modulus supplied with the prestressing steel coil. For ease of calculation, the elongations can be computed with an analysis software package, such as Sofistik or an excel based spreadsheet. These tools require the accurate geometric setting out coordinates of the duct. The theoretical elongations calculated are the "face to face" elongations, which is the elongation of the tendon between the anchorages.

Measuring of the actual tendon elongation

Prior to preforming the tensioning operation, the site team should calculate the required pressure in the jack to produce the correct tension force in the tendon. A rough calculation can be done to estimate the pressure in the jack by dividing the target force by the jack's piston area as shown in Equation 7. There is however friction in the jack which must be accounted for and therefore this calculation will always produce a lower pressure than desired.

$$P_{jack} = {P_0}\big/{A_{piston}} \tag{7}$$

P_{jack} = 6486 kN / 1237.01 cm² = 52.433 MPa

Instead, the supplier of the jack should provide a calibration certificate of the jack which shows the correlation between the pressure in the jack and a calibrated load cell. When calibrating the jack used for the Ashton Bridge, it was found that there is a good linear relationship between the internal jack pressure and an external load cell as shown in Figure 44.

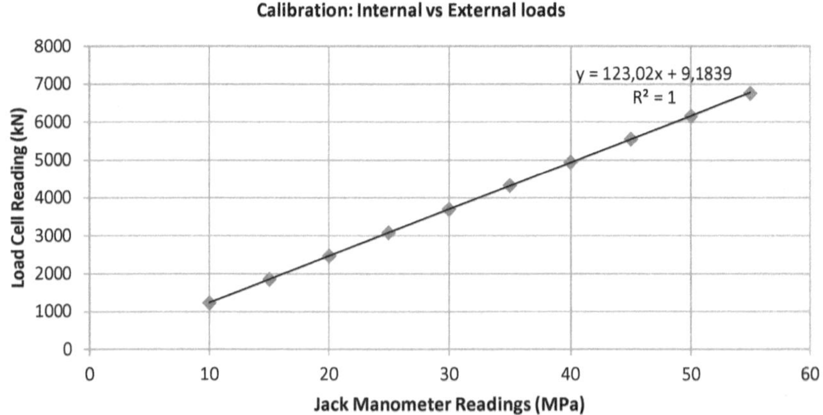

Figure 44 Internal vs external loads

The actual jack force applied to the structure was determined by using the linear regression equation in Figure 44, and then making the force the subject of the equation.

$$P_{jack} = (6486 - 9.1839)/123.02 = 52.6\ MPa \qquad (8)$$

Once the jack pressure has been confirmed, tensioning of the strands may commence. The site team must ensure that all wedges are seated and engaged on the near and far end of the deck. The jack is then fitted onto the back of the anchor plate and tensioning starts with the jack operator increasing the tension in the jack to 10 MPa before any measurements are taken. This is done so that any slack in the tendon is removed before any measurements are taken.

Tensioning then continues and the jack pressure is increased in 5 MPa increments, with the elongation of the jack recorded each time as shown in Figure 45. The operator and supervisor closely monitor the rate of elongation per 5 MPa increment to so ensure that no unpredicted behaviour takes place.

For the Ashton Arch Bridge, the expected elongation for the tie-beam is approximately 700 mm, which is appreciably high. The 6800 HoZ jack only has a piston stroke length of 300 mm, which means that the tendon is elongated to a length of about 300 mm before the jack has to seat the wedges in the anchor head and retract the piston so that the jack can perform a second and then finally a third stroke. The concept of a stroke is shown visually in Figure 45.

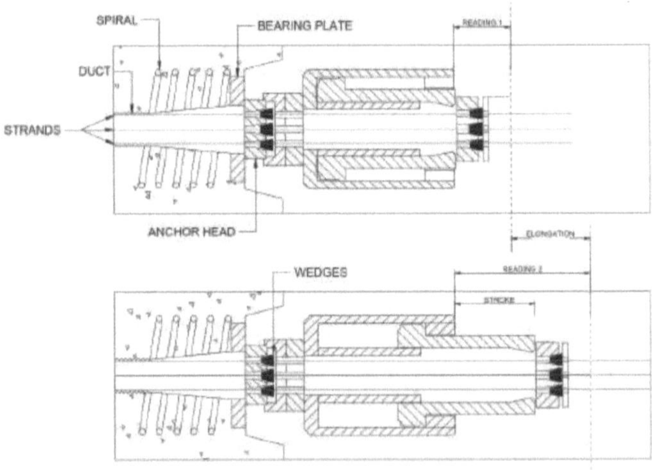

Figure 45 Physical elongation measurements

The pressure-displacement diagrams generated from plotting the data will always be linear for the pull from Jack Point 1. Since the first elongation measurement is taken after the 10 MPa mark, the elongation of the tendon prior to 10 MPa can be obtained by fitting a linear-regression line though the graph. The total physical elongation is then the portion of the regression line smaller than 0, plus the physically measured elongation. This concept is illustrated with Figure 46, where the linear regression line is the dotted line.

The pressure-displacement diagram from Jack Point 2 is however a non-linear graph, as the tendon is already taut before tensioning commences. As tensioning starts on the JP2, there is very little elongation measured behind the jack. The jack builds up pressure up until the pressure in the jack is equal to the pressure in the tendon. The little elongation measured up to this point is the elongation of the short length of tendon between the live anchor head and the wedges in the jack. When the jack pressure eventually equals and exceeds the force in the tendon, the wedges inside the live anchor head disengages and the tendon length becomes active and starts to elongate at a much greater rate than before. The total physical elongation is then the full measurement taken during the tensioning at Jack Point 2.

Typical graphs for pressure versus elongation are shown for tensioning from Jack Point 1 and tensioning from Jack Point 2 in Figure 46.

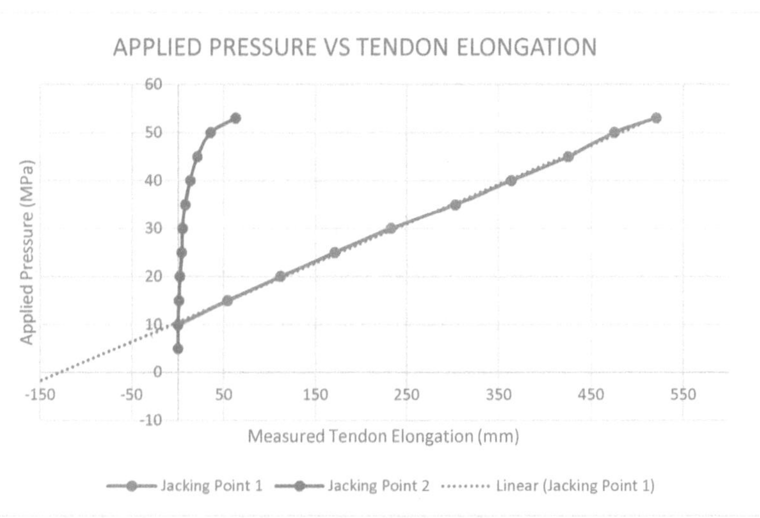

Figure 46 Typical Pressure vs Elongation diagram

The total physical elongation measurements can however not be used directly to compare with the theoretically calculated elongations. These elongations are made up of a number of components, as describe in Section 2.4.3, which must be removed before comparing with the theoretical elongations. This includes:

- The "face to face" elongation of the steel between the anchor plates.
- Elastic shortening of the concrete
- Elongation in the jack
- Draw in of the dead anchor

The elastic shortening is usually very small and is therefore regarded as negligible. The elongation of the tendon in the jack is estimated with the axial stiffness formulae $\Delta l = PL_{jack}/A_sE_s$. The portion of the steel strand in the 6800 Hoz Jack, L_{jack}, has been measured to be 640 mm. The standard draw-in of the wedges at the dead-end anchor is provided in the manufacturer's catalogue, and for this specific system the value used is 2 mm. This value must however be confirmed during tensioning, by measuring the physical draw in at the dead anchor after completion of the tensioning.

Finally, the face to face elongation may be calculated by subtracting the components mentioned above from the total elongation measured behind the jack. This face to face elongation is then used for comparison with the theoretically calculated elongation and therefore the validation of the post-tensioning force applied to the structure is achieved.

Friction coefficient calibration

If there is a difference between the theoretical tendon elongations and the physically measured elongation, the friction coefficients may be calibrated by suitable methods. Leonhardt (1964) proposed a simple method were the average wobble factor may be determined for a constant friction μ coefficient. This is achieved by calculating theoretical force-extension diagrams for varying wobble coefficients ranging between 0.0025 and 0.0065 rad/m, which are the boundary values in literature. The resulting diagram will resemble a set of radiating lines as shown in Figure 47. The field data may then be plotted on as scatter points on the diagram to determine the actual wobble value.

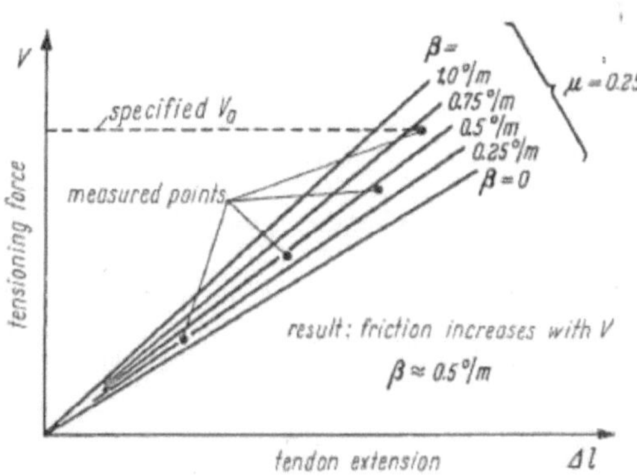

Figure 47 Wobble coefficient calibration (Leonhardt, 1964)

4.2.2. Tendon Lift off Testing

During post-tensioning it is necessary to check tendon elongations to verify that the correct friction loss assumptions were made during the design. If the actual measured elongations are significantly different from the design assumptions during tensioning from Jack Point 1, it may be necessary to determine the actual force losses by measuring the force at the anchor head at the Jack Point 2 with a lift-off test. The theoretical force which should be achieved may be read off the force diagram shown in Figure 48, located at Jack Point 2 and circled in red.

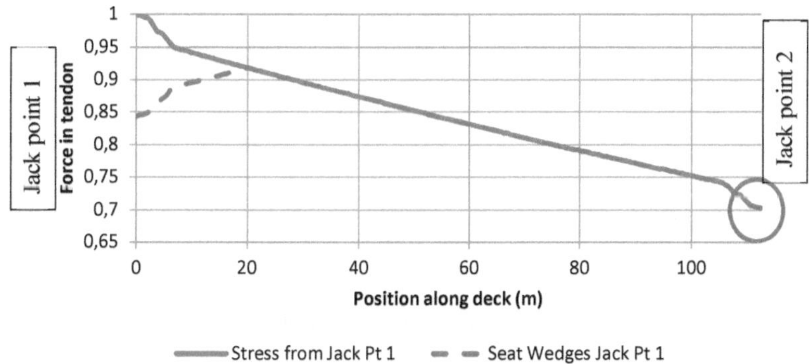

Figure 48 Normalised tendon-force profile after jacking from JP 1

The lift-off testing done at the Ashton Arch bridge was combined with the permanent tensioning at the second jacking point. Therefore it was performed without a stool and the jack was mounted directly against the anchor head as shown in Figure 45. When tensioning from the second Jack Point of a two-stage post-tensioning operation, the extension graph is similar to the graphs shown in Figure 26 and Figure 27. The difference between this method and the conventional lift-off test is that in this method the wedges in the anchor head will dislodge.

The test starts by increasing the load while measuring the extension of the jack accurately with a precision dial gauge (Figure 49). The jack pressure is increased in 5 MPa increments until 25 MPa, which is about 5 to 10 MPa below where lift-off is expected. At this stage the tendon is already highly tensioned and therefore little extension is visible on the dial gauge. After reaching 25MPa the jack pressure is increased in smaller increments of 1 or 2 MPa increments so that a precise force extension curve can be plotted. The force in the jack increases until it equals the force in the tendon at the anchor, plus any additional force required to overcome the friction of the wedges. This force in the wedges may be significant and may be evident by a sudden drop in jack pressure after the wedges have been dislodged (South African National Roads Authority Limited, 2011). A longer extension for each stress increment is noted thereafter and the audible clicking noise of the wedges releasing from the clips are heard. After reaching lift-off, stress increments are recorded at 5 MPa increments again and the tendon is tensioned to its design tensioning force and locked off.

Martin (South African National Roads Authority Limited, 2011) points out that after lift-off has taken place in the second stage, the graph will no longer be linear. This is because the length of tendon affected by the jack gradually increases from the point of the jack towards the other end of the tendon. The second stage of test curve is therefore a convex curve.

Figure 49 Jack with magnetic precision Dial Gauge

4.3. Structural Response Validation

4.3.1. Concrete strain measurement

The tied-arch bridge was originally modelled using Bentley's RM Bridge. A design review was later conducted by the Consulting Engineer's UK long-span bridge specialist team using SOFISTIK. These specialist bridge analysis software packages were used for the analysis of the structure's behaviour during construction and during its service life. For the purpose of this book, the stress state of the tie-beam after tensioning of each tendon will be extracted from the FEM model and tabulated as the theoretical behaviour. This will be six stress values per Northern tie-beam and six stress values per Southern tie-beam, i.e. one value per tendon tensioned. The stress values were extracted at the position when the strain gauges are located, namely 8 metres form the edge of the deck adjacent to the arch spring point. The stress values are primarily made up of axial forces and the small bending shown in the model is regarded as negligible.

The FEM models analyse tensioning from Jack Point 1 and Jack Point 2 as a single load case. Therefore, there are only 12 load cases for tensioning the tie-beams, and not 24 as in reality. This is however adequate for comparison with actual deformations of the strain gauges, as the strain gauges are located close to edge of the deck on the Montague side (East) and will feel minimal effects from tensioning from the Robertson side (West).

For the actual behaviour of the structure, a resistance based instrumentation system was chosen for measuring strain on the structure. Resistance type strain gauges have been imbedded in the bridge at critical locations where maximum stresses are anticipated. For the tie-beams, strain gauges are located at the connection between the tie-beam and spring point, as it is expected that the largest stresses will occur here when the bridge is subject to the maximum service load. At each critical section of the tie-beam, five strain gauges have been installed at different heights so that the axial and bending deformation can be measured. It was also anticipated that there will be failure of some of the strain gauges, therefore it was useful to have redundant gauges. These gauges are shown in Figure 50 and Figure 51 and have been labelled SC1Mod2_ai1 to SC1Mod2_ai5, which corresponds to the input channel of the data acquisition system.

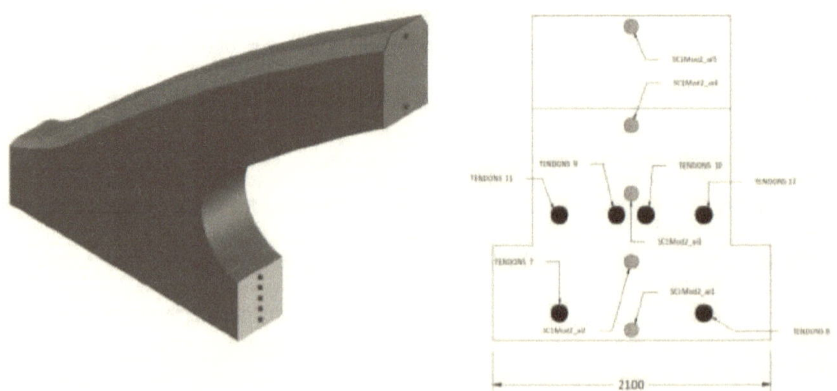

Figure 50 Arrangement of dynamic strain gauges in cross-section (Northern Arch)

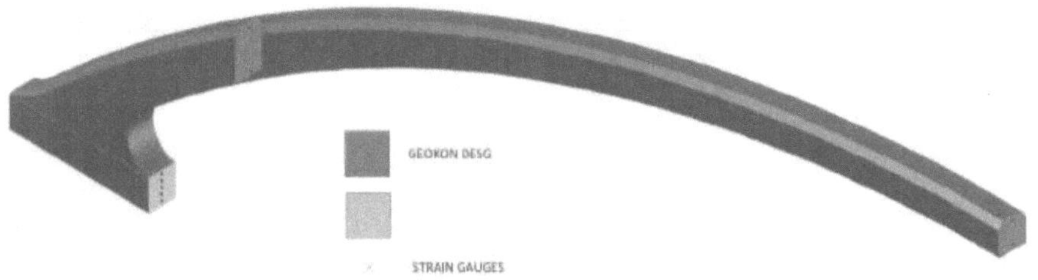

Figure 51 Arrangement of dynamic strain gauges (Southern Arch)

The arch ribs have also been instrumented with strain gauges, but the analysis of these members are beyond the scope of this book. It was intended that robust Geokon instruments be installed throughout the entire structure, but due to the long lead time required to procure them, they were only used for the arch ribs. The tie-beams, which were constructed first, were installed with foil-type resistance strain gauges, which was sourced locally and assembled at the university laboratory. Each foil-type strain gauge was cemented onto a 1 metre long Y16 reinforcement bar and protected with a 25 x 25 mm aluminium angle with silicone smeared along all edges to prevent moisture ingress during concreting (Figure 52).

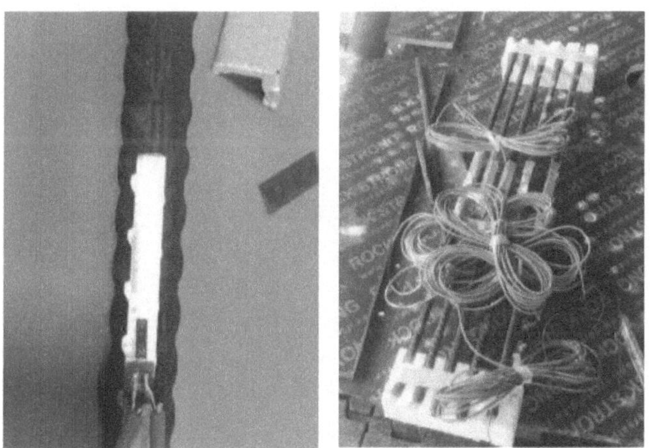

Figure 52 Manufacture of rebar mounted foil type strain gauges

The uniaxial foil strain gauges shown in Figure 53 have a 120 Ω resistance, which is compatible with the data logging equipment provided by the University of Cape Town. The gauge is 5 mm long and was also chosen to be compatible with the thermal expansion of reinforcement steel. A 20 mm gauge length would have been preferred for ease of mounting onto the reinforcement, but no stock was available at the time.

Figure 53 Kyowa KFGS-5-120-C1 strain gauge

The strain gauges feed into a Universal Strain Terminal Block (SCXI – 1314), which has 8 channels and can therefore read 8 strain gauges simultaneously. Each channel has a connection for the quarter-bridge completion resistor, i.e. the foil-type strain gauge. The strain gauge is connected to the terminal block to create a Quarter-Bridge Type 1 Connection, which is illustrated in Figure 54. Cables are connected to the positive excitation signal (P+), positive input signal (S+) and quarter-bridge completion resistor connection (QTR).

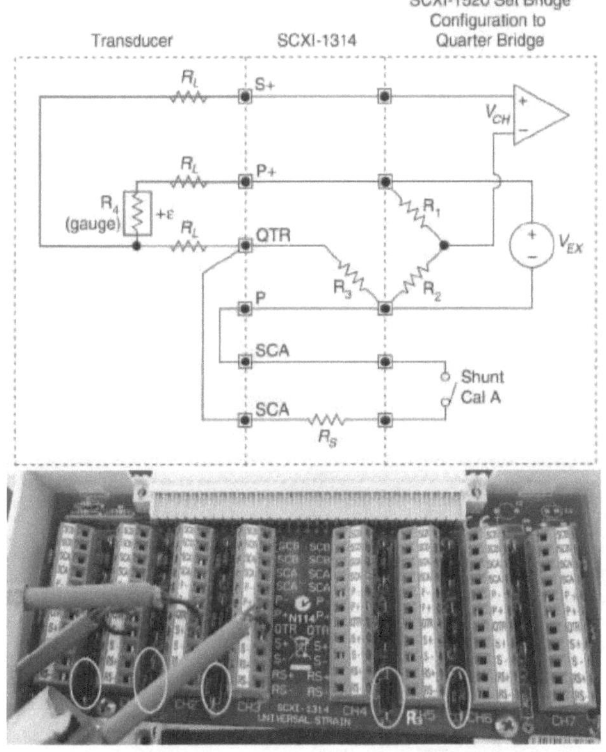

Figure 54 Quarter Bridge 1 Circuit Diagram and actual board of SCXI 1314

The symbols referred to in the circuit diagram of Figure 54 are as follows:

- R_1 & R_2 are half-bridge completion resistors   (In the SXCI-1540 module)

- R_3 is a quarter bridge completion resistor   (in the SXCI-1314 module)

- R_4 is the strain gauge   (in the concrete structure)

- R_L is the resistance in the cables   (Between concrete and SXCI-1314 module)

- V_{EX} is the excitation voltage   (In the SXCI-1540 module)

- V_{CH} is the measure voltage   (In the SXCI-1540 module)

The Terminal Block (SCXI-1314) is in turn mounted to the front of the SCXI-1520 module. The SCXI-150 module is housed in the SCXI chassis box and contains the remaining resistors which make up the quarter-bridge strain gauge. It also controls the excitation and measurement of voltage signals, which is forwarded to the computer for storage and processing. The NI data capturing equipment and the setup onsite is shown in Figure 55.

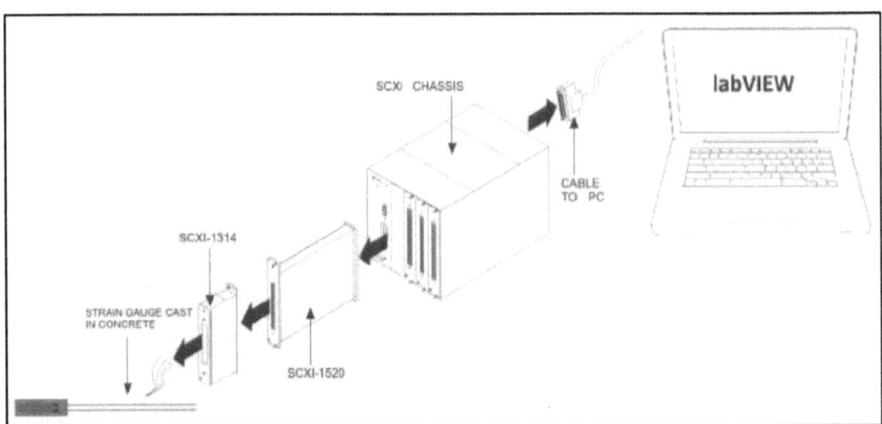

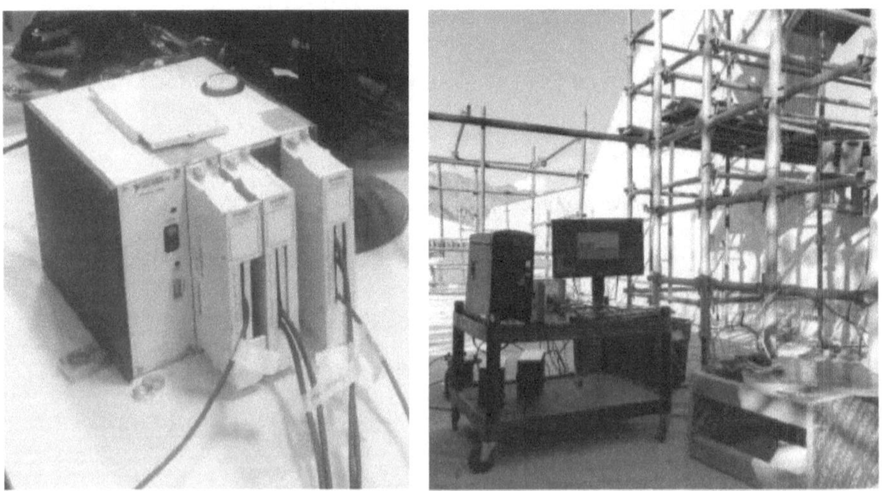

Figure 55 NI Data capturing equipment

Strain measurement took place over a period of 6 days as the post-tensioning activity was taking place. Strain measuring started before the post-tensioning of a tendon began and was stopped after the post-tensioning ended. There was a total of 12 tendons that were tensioned in the tie-beam, where each tendon was tensioned once from each Jack Point, therefore 24 post-tensioning events took place. Data was lost for 7 of these 24 post-tensioning events. Also to note is that there were 10 strain gauges installed in the tie-beams, 5 in each of the two tie-beams. Three strain gauges malfunctioned prior to starting the post-tensioning measuring and therefore only 7 strain reading are reported in this book. Of the 7 strain gauges only 3 reported accurate measurements throughout the post-tensioning of the tie-beams. The other gauges appear to have lost the strain history when the strain measuring restarted after a break.

The post-tensioning activity strains were measured at a frequency of 16Hz, but sampling size was reduced during processing so that the large data pool could be processed with the available software. Data was captured on a computer in a *.txt file where it was transferred and processed with Microsoft Excel. Managing large data in Excel is problematic and therefore the data sample size was reduced to a sampling rate of 1 reading per second. This was done at random by using the V-lookup function, where the 1st occurring data cell in a second period was used.

The strain points for each post-tensioning exercise was plotted to a graph and a "moving average" trendline was fitted to the data. Because of the high scatter strain points, it was decided that a period of 255 was to be used as the Period of the regression line. The strain data scatter over time during tensioning can be found in Appendix B.

For comparing the theoretical data with the actual data, the actual data was further refined. An average value over a period of two minutes was taken before and after each tensioning operation. These strain values were then tabulated and converted to equivalent stress form, assuming a elasitc-modulus of 31.5 GPa.

4.3.2. Longitudinal deck shortening

The theoretical deck shortening can be obtained in two ways. A more accurate displacement of the deck can be extracted from the Designer's FEM model for validation. An additional check can be done with a hand calculation to check the FEM results.

For calculating the as-built deformation, four stainless steel pegs were installed on each corner of the bridge, behind the spring point and on-top of the tie-beams. The four points were measured immediately the before application of the post-tensioning load and immediately after completion, so that the elastic shortening of the deck could be captured. Readings were then also taken at regular intervals after the completion of post-tensioning to measure the resulting creep shortening effects.

The pegs can easily be accessed throughout the construction period for measurements of the movement of the deck during important construction events. Results obtained by measuring the shortening of the deck can be used to verify that the anticipated elastic shortening of the deck is as anticipated and that there is no restraint to shortening of the deck.

A Trimble M3 Total Station was used to measure the coordinate and a typical total station can measure distances with an accuracy of about 1.5 mm and 2 parts per million (ppm) over distances up to 1500 m. A Leica sprinter electronic dumpy level was used to pick up levels and is accurate to 1.5mm.

5. VALIDATION OF TEST RESULTS

The structural behaviour will be validated by evaluating the input and output parameters, by comparing theoretically calculated results with field measured results. The field measured results are presented in this section.

5.1. Load Validation

The input parameter refers to the post-tensioning load applied to the structure, which will be validated by tendon elongation measurements and tendon lift-off testing.

5.1.1. Tendon elongation measurement

The "Theoretical Elongations" were calculated before commencing with the tensioning operation onsite, using the theoretical friction and wobble coefficients. These theoretical calculations were checked by the contractor and the designer, so that a consensus was reached. The friction factor $\mu = 0.25$ and wobble factor $k = 0.00025$ rad/m was agreed on, as this matched the proprietary catalogue of the prestressing system manufacturer for the specific strand and duct used. The values are regarded as the base friction coefficients. The theoretical "face to face" elongations are the same for ducts with the same geometry and the values calculated are summarised in Table 3.

Table 3 Base theoretical elongations

Tendon Number	Theoretical Elongation (mm)		
	Jack Point 1	Jack Point 2	Total
Tendon 1, 2, 7, 8	656	55.5	711
Tendon 3, 4, 9, 10	659	56.0	715
Tendon 5, 6, 11, 12	647	56.5	703

The theoretical elongations are adjusted from the design elastic-modulus value of 195 GPa to the elastic-modulus tagged on the coil of strand delivered to site. An elastic-modulus can range between 185 GPa to 205 GPa, meaning that the theoretical elongation can vary by as much as ±5% because of the e-modulus adjustment.

The physically measured elongations are directly measured from the jack during tensioning and compared to the adjusted theoretical elongation. A summary of the results of field measured elongations versus the theoretical elongations are shown in Table 4 and Table 5.

Table 4 summarises the results from tensioning from Jack Point 1 and Table 5 summarises the results from tensioning from Jack Point 2. When tensioning from Jack Point 1 the average elongations on the Northern tie-beam (i.e. Tendons 1 - Tendon 6) was found to be -6.0% under-extended and the Southern tie-beam (i.e. Tendon 7 – Tendons 12) was found to be -6.8% under-extended. The outliers identified were Tendon 2, Tendon 7 and Tendon 10 that were -10.0 %, -16.7 % and -17.3 % respectively. A negative percentage difference indicated that the extension was less than anticipated and therefore an under-extension.

Table 4 Elongation results after tensioning from Jack Point 1

Tendon Number	Field Measured Elongation	Design Elongation (mm)	Adjusted Design Elongation	Percentage Difference (%)	Average (%)
Tendon 1	619	656	647	-4.2	
Tendon 2	582	656	650	-10.4	
Tendon 3	587	659	646	-9.0	-6.0
Tendon 4	625	659	659	-5.1	
Tendon 5	651	647	659	-1.1	
Tendon 6	610	647	648	-5.9	
Tendon 7	534	656	641	-16.7	
Tendon 8	598	656	635	-5.8	
Tendon 9	612	659	659	-7.1	-6.8
Tendon 10	534	659	645	-17.3	
Tendon 11	640	647	624	2.5	
Tendon 12	640	647	617	3.8	
					-6.4

The elongations measured at Jack Point 2 were more variable than those at Jack Point 1. This is primarily because any under-extension encountered at Jack Point 1 would be corrected at

Jack Point 2. This was most obvious for Tendon 2, Tendon 7 and Tendon 10 that was significantly under-extended during tensioning from the 1st Jack Point and therefore apparent over-extension was witnessed when tensioning from Jack Point 2.

Since the target design elongation for Jack Point 2 is very low relative to what is expected for elongations at Jack Point 1, any additional slack picked up due to an under-extension at Jack Point 1 will have a significant effect on the percentage difference. This is apparent in Table 5.

Table 5 Elongation results after tensioning from Jack Point 2

Tendon Number	Final Measured Elongation	Design Elongation (mm)	Adjusted Design Elongation	Percentage Difference (%)	Average (%)
Tendon 1	73.7	55.5	54.8	34.6	**38.56**
Tendon 2	111.2	55.5	55.0	102.1	
Tendon 3	68.3	56.0	54.9	24.4	
Tendon 4	71.1	56.0	56.0	27.0	
Tendon 5	62.0	56.5	57.5	7.7	
Tendon 6	76.6	56.5	56.6	35.3	
Tendon 7	147.3	55.5	54.2	171.5	**40.37**
Tendon 8	66.9	55.5	53.8	24.4	
Tendon 9	60.1	56.0	56.0	7.3	
Tendon 10	99.3	56.0	54.9	81.0	
Tendon 11	31.9	56.5	54.5	-41.4	
Tendon 12	53.5	56.5	53.9	-0.6	
					39.47

When summing the extensions of Jack Point 1 and Jack Point 2, to compare with the total design elongations it was found that the under-extensions encountered at Jack Point 1 was balanced with the over-extensions encountered at Jack Point 2. The second pull, from Jack Point 2, improved the results significantly so that the total extensions was with-in specification for the design friction and wobble coefficients. Tendon 3 was marginally under-extended and Tendon 10 was more significantly under-extended. The average for the Northern and Southern tie-beam tendons were however within 3 % for each group and therefore comply with COLTO specification. The summary of the total extensions, with the base friction coefficients are is shown in Table 6.

Table 6 Total elongation results after tensioning from Jack Point 1 & 2

Tendon Number	Field Measured Elongation	Design Elongation (mm)	Adjusted Design Elongation	Percentage Difference (%)	Average (%)
Tendon 1	695	711	702	-0.9	
Tendon 2	695	711	705	-1.3	
Tendon 3	657	716	702	-6.3	-2.3
Tendon 4	698	716	716	-2.5	
Tendon 5	715	704	716	-0.2	
Tendon 6	688	704	705	-2.4	
Tendon 7	683	711	695	-1.7	
Tendon 8	667	711	689	-3.1	
Tendon 9	674	716	716	-5.9	-2.9
Tendon 10	635	716	701	-9.5	
Tendon 11	674	704	679	-0.8	
Tendon 12	695	704	671	3.7	
					-2.6

Friction coefficient calibration

Since the under-extensions were consistently lower for the first pull, it was deemed necessary to calibrate the friction coefficients, so that the additional friction losses in the FEM model could be accounted for. This *model update* was necessary so that the *output* structural response could be validated as well. The friction factor μ was adjusted from 0.25 to 0.3 to account for additional light rust on the interface of the strand and duct.

The wobble coefficient was calibrated using the method described in Section 4.2.1. The radiating lines and field data scatter is shown in Figure 56. From the plotted data, the calibrated wobble factor k was taken to be 0.0045 rad/m for Duct 1, 2, 7, 8 and Duct 3, 4, 9, 10. Similarly the calibrated wobble factor k was taken to be 0.0035 rad/m for Duct 5, 6, 11, 12. Field data points of Tendon 7 and Tendon 10 were regarded as outliers and were plotted with triangles, but were not considered in determining the calibrated wobble factor k.

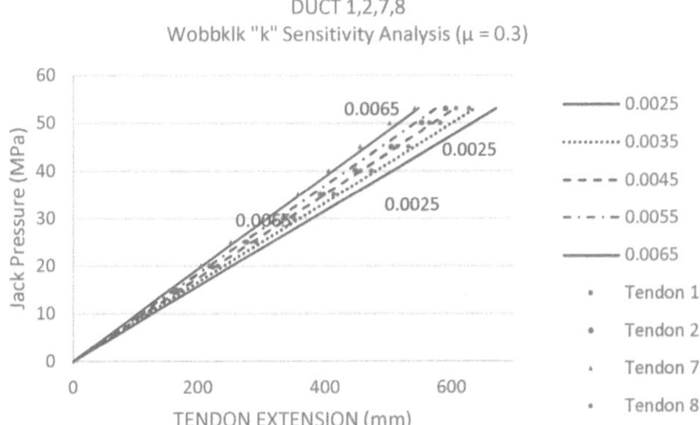

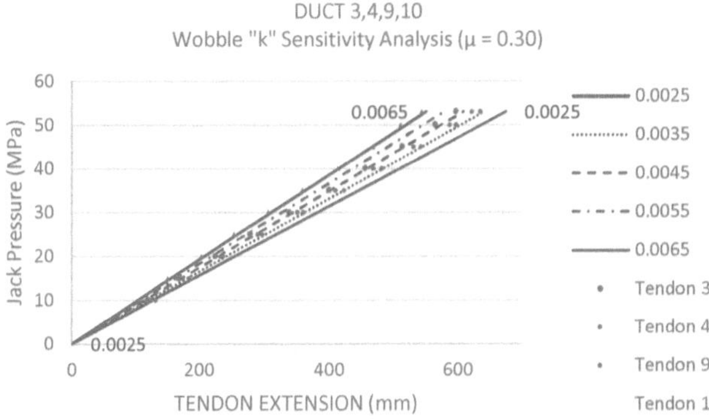

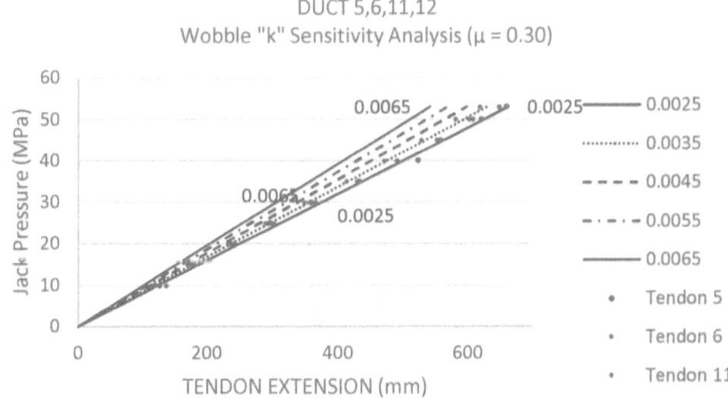

Figure 56 Sensitivity Analysis of Wobble factor k

The theoretical extensions were then re-calculated with the calibrated wobble factors and compared to the actual extensions. The comparison is tabulated in Table 7, which reflects a closer correlation with the physically measured elongations.

Table 7 Total elongation validations with updated friction coefficients

Calibrated wobble k (rad/m)		WITH CALIBRATED WOBBLE K VALUE									
		SITE MEASURED DATA			THEORETICAL		ADJUSTED THEORETICAL			% DIFFERENCE		
		PT1	PT2	TOTAL	PT1	PT2	PT1	PT2	TOTAL	PT1	PT2	TOTAL
Tendon 1	0.0045	619	76	695	606	77	598	76	673	3.6	0.0	3.2
Tendon 2	0.0045	582	113	695	606	77	600	76	676	-3.0	48.7	2.8
Tendon 3	0.0045	587	70	657	609	77	596	76	672	-1.6	-7.1	-2.2
Tendon 4	0.0045	625	73	698	609	77	609	77	686	2.7	-5.3	1.8
Tendon 5	0.0035	651	64	715	630	64	641	65	706	1.6	-1.8	1.3
Tendon 6	0.0035	610	79	688	630	64	631	64	695	-3.3	22.6	-0.9
Tendon 7	0.0045	534	149	683	606	77	592	75	667	-9.9	98.9	2.4
Tendon 8	0.0045	598	69	667	606	77	587	74	661	2.0	-7.4	0.9
Tendon 9	0.0045	612	62	674	609	77	609	77	686	0.5	-19.5	-1.7
Tendon 10	0.0045	534	101	635	609	77	596	76	672	-10.5	33.9	-5.5
Tendon 11	0.0035	640	34	674	630	64	608	62	669	5.3	-45.1	0.7
Tendon 12	0.0035	640	55	695	630	64	600	61	661	6.6	-9.0	5.2
									Average	**-0.5**	**9.1**	**0.7**

The average difference between the updated theoretical elongations and the site measured elongations is +0.7%. This validates the predicted tension modelled in the tendons. The site measure elongations and the pressure-displacement diagrams are added to Appendix C of this book.

5.1.2. Tendon lift-off testing

Theoretical forces which were expected at Jack Point 2 were calculated from tendon-force profile diagrams with calibrated friction coefficients. Lift-off testing was conducted after tensioning a tendon from Jack Point 1, immediately before tensioning the same tendon from Jack Point 2. The lift-off testing may be regarded as part of the Jack Point 2 tensioning procedure. The resulting pressure-elongation diagrams were used to determine the lift-off force values and can be found in Appendix D. The theoretical and site measured results have been extracted from theses diagrams and are tabulated in Table 8.

Table 8 Lift-off Test Validation

Cable Number	Theoretical Force at Jack Point 2 (kN)	Gauge Pressure (MPa)	Measured Force (kN)	Percentage Difference (%)
Tendon 1	3496	29	3580	2.4
Tendon 2	3496	31.5	3889	11.2
Tendon 3	3540	28.5	3517	-0.6
Tendon 4	3540	32	3951	11.6
Tendon 5	3814	36	4321	13.3
Tendon 6	3814	33.5	4137	8.5
Tendon 7	3496	29.4	3630	3.8
Tendon 8	3496	28.5	3517	0.6
~~Tendon 9~~	~~3540~~	~~37~~	~~4568~~	~~29.0~~
Tendon 10	3540	26.5	3272	-7.6
Tendon 11	3814	35	4322	13.3
Tendon 12	3814	33.5	4137	8.5
			Average:	<u>5.9</u>

The results indicated a tendency for the measured force at Jack Point 2 to be higher than the expected theoretical forces. This is likely because of the additional force required to overcome the friction in the wedges. Tendon 9 has been identified as an outlier, since the force required to overcome the friction in the wedges is noted to be significantly higher, and it is evident when looking at the pressure-elongation diagram in Figure 57. This value has therefore been omitted when calculating the average of the percentage difference, which is equal to 5.9%.

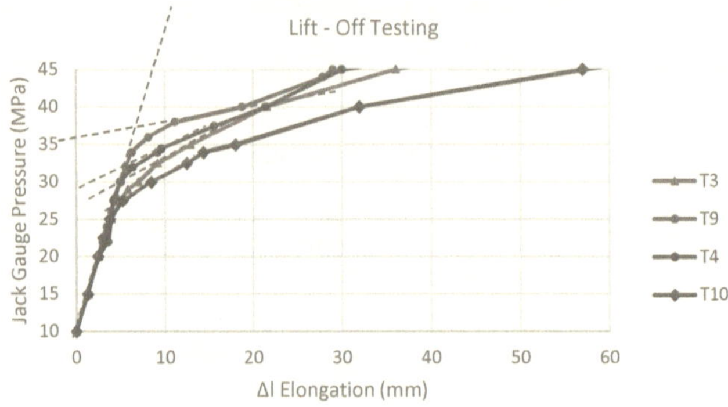

Figure 57 Typical lift off test results

5.1.3. Result comparison

Both the tendon elongation test and the lift-off test are measures for validating the force in the

tendon. Since the results of the tendon elongations are closer to the predicted results and are

not affected by wedge friction, there is a higher confidence that the tendon elongation results

match the actual force state of the tendons. The tendon elongation is directly proportional to

the force in the tendon as shown in Equation 4 and therefore the percentage difference may be

compared with each other as shown in Figure 58.

Percentage Difference (%)		
Cable Number	Lift-off Force	Tendon Elongation
Tendon 1	2.4	3.6
Tendon 2	11.2	-3.0
Tendon 3	-0.6	-1.6
Tendon 4	11.6	2.7
Tendon 5	13.3	1.6
Tendon 6	8.5	-3.3
Tendon 7	3.8	-9.9
Tendon 8	0.6	2.0
Tendon 9	29.0	0.5
Tendon 10	-7.6	-10.5
Tendon 11	13.3	5.3
Tendon 12	8.5	6.6
average	5.9	-0.5

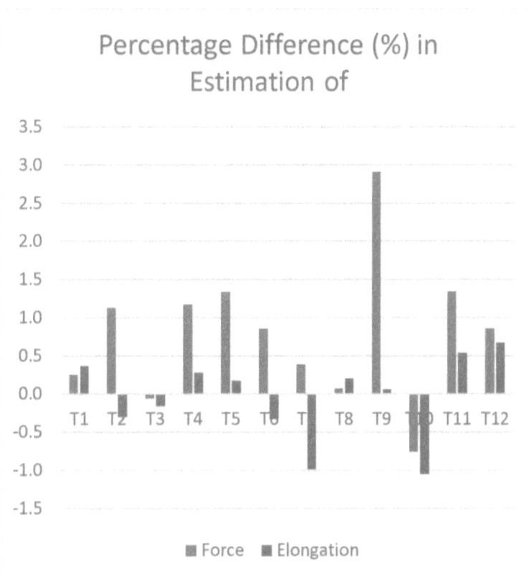

Figure 58 Load % difference comparison

Figure 58 visually illustrates the lower average prediction of the tendon elongations and the over-estimation of the lift-off measurements.

5.2. Structural Response Validation

The FEM model was updated with the validated friction coefficients prior to calculating the structural response which is presented in this section. These structural responses are the strain response and the displacement response.

5.2.1. Concrete strain measurement

The theoretical calculation of the stress state of the Tie-Beam after the post-tensioning construction stage was extracted from the FEM models. Each tendon tensioned was modelled discretely. The resulting force in each tendon after the post-tensioning stage is shown in Figure 59. The forces and stresses resulting in the concrete, at the position where strain gauges have been installed, are shown in Table 9. The value of these forces varies from the force in the tendon due to the eccentricity of the load relative to the neutral axis of the tie-beam and the distribution of the load to the adjacent longitudinal beams.

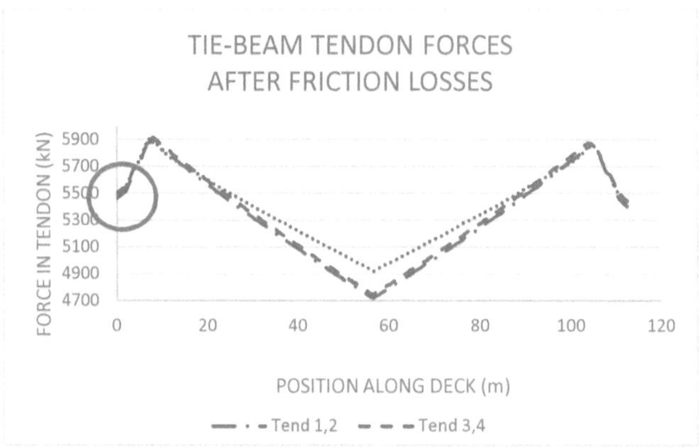

Figure 59 - Tendon Forces after Friction Loss

The "discrete" labelled column in Table 9 shows the effect of tensioning a single tendon in a tie-beam. From the analysis results it is clear that the discrete load adds compression to the tie-beam being tensioned and reduces compression in the tie-beam on the opposite side of the deck. This is graphically represented by the outward bowing of the deck in the FEM model, for a *difference load case* shown in Figure 60 . The *difference load case* shows the effect of the discrete load on the deck, ignoring the stress history of the structure and therefore the graphic is exaggerated.

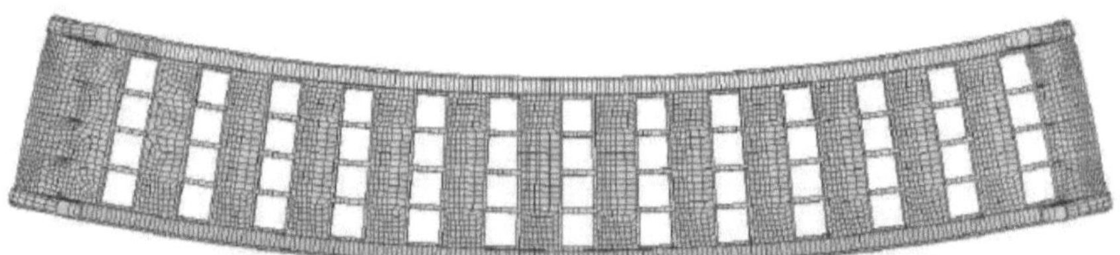

Figure 60 Lateral bowing of deck due to post-tensioning

The cumulative columns in Table 9 shows the build-up of compression in each tie-beam after each consecutive tendon tensioning. This is also plotted on the graph in Figure 61 and Figure 62 as the dotted red theoretical line. The total compression built-up in each tie-beam is approximately 29967 kN, which is 77 % of initial jacking force P_0. The remainder of the force was lost to elastic shortening of the deck, wedge slip during lock-off and force distribution among the less-stiff intermediate longitudinal beams.

Table 9 Force in Concrete due to Post-tensioning

Sequence	TENDON	FORCE (kN)				Stress (MPa)	
		Discrete		Cumulative		Cumulative	
		North TB	South TB	North TB	South TB	North TB	South TB
1	3	-5450	380	-5450	380	1.84	-0.13
2	10	380	-5450	-5070	-5070	1.71	1.71
3	9	316	-5320	-4754	-10390	1.60	3.50
4	4	-5320	316	-10074	-10074	3.39	3.39
5	2	-5130	205	-15204	-9869	5.12	3.32
6	7	205	-5130	-14999	-14999	5.05	5.05
7	12	482	-5600	-14517	-20599	4.89	6.94
8	5	-5600	482	-20117	-20117	6.77	6.77
9	8	485	-5380	-19632	-25497	6.61	8.58
10	1	-5380	485	-25012	-25012	8.42	8.42
11	11	195	-5150	-24817	-30162	8.36	10.16
12	6	-5150	195	-29967	-29967	10.09	10.09

The physical strain deformations were measured continuously during post-tensioning events and includes results from tensioning both Jack Point 1 and Jack Point 2. Graphs of the continuously measured data can be found in Appendix B, which includes a summarised table of important strain readings. This includes measurements of strain before and after each tensioning event. These strains were converted to equivalent stresses, using an elastic-modulus of 31.5 GPa for easier comprehension. The comparison is shown graphically in Figure 61 and Figure 62, where the measured data has been plotted with blue lines. Faulty strain gauge readings, where strain history is repeatedly lost, are plotted with grey lines.

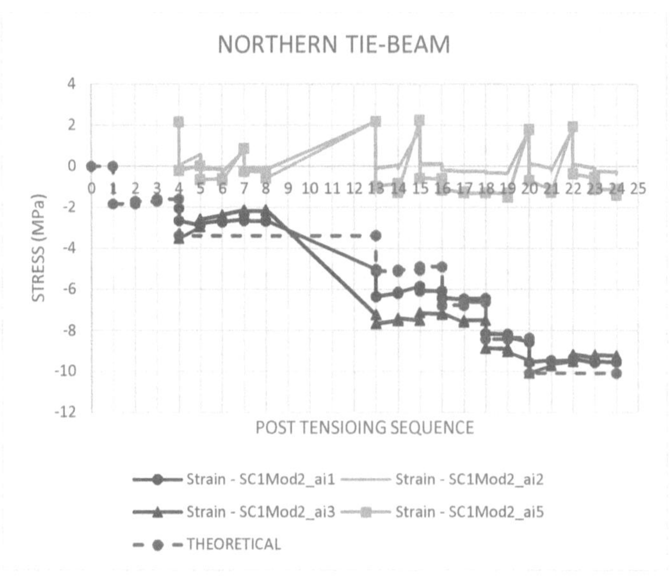

Figure 61 Stress validation for post-tensioning of the Northern Tie-Beam

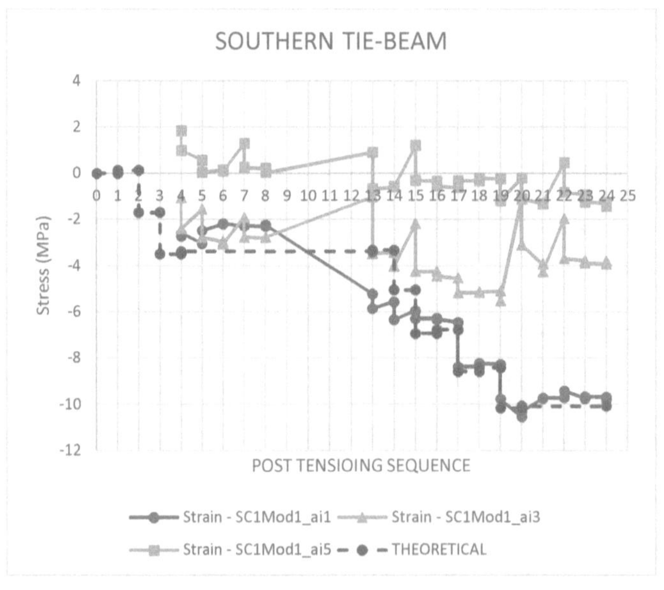

Figure 62 Stress validation for post-tensioning of Southern Tie-Beam

The results indicated that the strain gauge readings are very close to the theoretical compression values. A comparison of the final stress in the tie-beam after completion of the tensioning is tabulated in Table 10.

Table 10 Validation of strain measurement

Member	Theoretical (MPa)	Measured Strain (με)	Measured (MPa)	Percentage Difference (%)
Northern Tie Beam	10.1	298	9.4	-6.9
Southern Tie Beam	10.1	312	9.7	-4.0
			Average	-5.5

5.2.2. Longitudinal deck elastic shortening

The theoretical elastic deck shortening in the longitudinal direction was extracted from the finite element model and was also checked with a basic hand calculation. The finite element model indicates that the expected shortening of the deck due to tie-beam post-tensioning is equal to 28 mm. The cumulative axial compressive force in the tie-beam at the end of the post-tensioning is 29967 kN. A hand calculation check was done using Equation 6, where:

$$P = 29\,967 \text{ kN} \qquad L = 112.5 \text{ m} \qquad A = 2.97 \text{ m}^2 \qquad E = 31.5 \text{ GPa}$$

$$\Delta l = \frac{(PL)}{(A_c . E_c)} = \frac{(29\,967 x 10^3 \; x \; 112.5)}{(2.97 \; x \; 31.5 x 10^9)} = 35.3 \; mm$$

The calculation is conservative, since it ignores the loss in compression force throughout the length of the beam. This value is therefore an upper bound order size check for the FEM calculation. The actual deformation of the deck was measured by site surveyors and is the difference between the coordinates of each corner of the bridge deck measured before and after post-tensioning of the tie-beams. An exaggerated illustration of the actual deck deformation is shown in Figure 63.

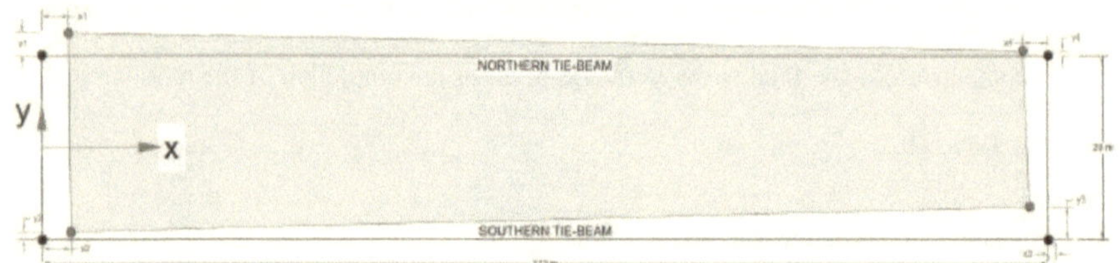

Figure 63 Exaggerated deflection of the deck due to Tie Beam Post-tensioning

The shortening of the deck in the longitudinal direction is 19 – (-15) = 34 mm on the Northern side of the deck and 20 – (-7) = 27 mm on the Southern side of the deck. The average deck shortening is therefore 30.5 mm. The displacement ordinates of each corner is summarised in Table 11, which should be read in combination with .Figure 63

Table 11 Deck displacement measurement

Point	X (mm)	Y (mm)
1	19	13
2	20	5
3	-7	18
4	-15	4

Compression of the concrete deck can only take place if the deck is not restrained during tensioning. The actual versus theoretical deck shortening is compared in Table 12 and shows a close comparison to what was predicted with the FEM model.

Table 12 Deck shortening validation

MEMEBER	Deck shortening (mm)			Equivalent Strain (με)	Percentage Difference (%)
	Theoretical Calculation	Actual measured	Hand Calculation		
Northern Tie-Beam	28	34	35	302	+21.4 %
Southern Tie-Beam	28	27	35	240	-3.6%

From the results it appears the Northern tie-beam has received higher compression than the Southern tie-beam. There is a notable 8 mm difference in the shortening measured on the Eastern side of the deck.   This variance in shortening may be partially attributed to the longitudinal friction resistance of the formwork which will be confirmed when hanger tensioning takes place and the deck is lifted up off the formwork.

Such a large variance is not desirable when attempting to validate a FEM model. This is because the stress experienced in the concrete is sensitive to the displacement measure. Here 1 mm of displacement is equivalent to 0.28 MPa. The total station machine used to measure the elongation has a tolerance of 1.5 mm. This excludes the human error in operating the machine and holding the surveyor prism which could be an addition 1.5 mm. Two measurements were taken, one on each side of the deck, to calculate the elastic shortening of the deck, which could possible double a measurement error.

5.2.3. Result comparison

Both the strain gauge measurements and the elastic shortening of the deck are strain responses of the deck due to post-tension loading and may therefore be compared with one another as a strain measurement as is done in Table 13.

Table 13 Strain response comparison

Member	Strain gauge measurement ($\mu\varepsilon$)	Strain due to elastic shortening ($\mu\varepsilon$)
Northern Tie-Beam	298	302
Southern Tie-Beam	312	240

It is important to note that the stain gauge readings were taken at a point in the beam where maximum compression is anticipated and therefore it should be higher than the strain due to elastic shortening. The strain due to elastic shortening of the deck is an average strain along the length of the deck. This implies that the 302 mm measurement is the outlier and this is confirmed when elauating Table 14, which indicates a higher deck shortening than predicted for the Northern tie-beam.

Table 14 Strain % difference comparison

Member	Percentage Difference (%)	
	Strain gauge measurement	Elastic shortening measurement
Northern Tie-Beam	-6.9	+21.4
Southern Tie-Beam	-4.0	-3.6

6. CONCLUSION

The *general objective* of the research presented in this book is to validate the structural behaviour of the tie-beams after the post-tensioning construction stage. This was done by comparing the theoretically calculated parameters with field measured tests results. Four different tests were undertaken to validate the load applied to the structure and the resulting response of the structure. This included tendon elongation measurement, lift-off testing, concrete strain measurement and deck shortening measurement.

The measuring of the tendon elongations revealed that there was an under estimation of the friction coefficient, which meant that the model predicted a higher force in the tendon than what was in reality. The friction coefficients were calibrated and updated in the FEM model, so that the force in the tendon was validated.

Tendon lift-off testing was undertaken as a secondary check of the force in the tendons. The results predicted a higher force in the tendon than was anticipated. This is probably due to the additional force required to overcome friction in the wedges when jacking from the second jacking point. This test method may be improved in future validations by making use of a stool, so that an accurate lift-off is achieved immediately when the anchor head lifts off the bearing plate as demonstrated in literature under Section 4.2.2.

Strain was measured at vulnerable zones with resistance type strain gauges but were converted to stress values for easier comprehension. The results showed a good correlation with theoretically calculated stresses, as the average percentage difference was - 6%. The negative sign indicating that the stress measured was lower than anticipated. Two factors may contribute to this discrepancy. The assumed elastic-modulus of the concrete and the assumed boundary conditions. The deck was supported on formwork at the time of post-tensioning which may restrain compression of the concrete during post-tensioning. The strain should be verified again during hanger post-tensioning, when the deck is lifted from the supporting formwork.

The elastic shortening of the deck was measured by surveyors by picking up coordinates on the corner of the deck before and after post-tensioning. The difference of the ordinates in the longitudinal direction was taken as the deck shortening. The shortening captured on the Northern tie-beam was 34 mm and the Southern tie-beam was 27 mm, whereas the FEM model predicted a shortening of 28mm. The excessive shortening noted on the Northern tie-beam may be due to measurement errors. The reading on the Southern tie-beam corresponds well with the theoretical calculations. The readings obtained from the measurements of deck shortening are therefore good for determining an order size comparison, but are too variable for validation of the FEM model. Future readings should therefore be done with instruments which are less prone to machine and human error and are accurate to at least 1 mm. This may take the form of measurement with a double graduated ruler or a dial gauge from a fixed reference point.

When comparing the results from all field tests, it may be concluded that the tendon elongations measurement was successful in validating the load applied to the structure. The concrete strain measurement was successful in validating the strain response of the structure. Model updating may be done to further refine the structural response of the FEM model to better correlate with the measured strain data. This may be done by updating the elastic modulus of the concrete or the boundary conditions of the structure.

The lift-off testing and deck-shortening methods show promise in predicting the structural behaviour, but must be improved so that it may be used to validate FEM models in the future. If these test methods are refined the validation process developed in this book may be used for validation of future post-tensioned concrete bridges.

7. REFERENCES

Bazant, Z., & Baweja, S. (1996). TC 107-GCS: GUIDELINES FOR THE FORMULATION OF CREEP AND SHRINKAGE PREDICTION MODELS. *Materials and Structures*, 587-593.

Berverly, P. (2013). *FIB Model Code for Concrete Strucutres 2010.* Berlin: Wilhelm Ernst & Sohn.

Breen, J., Davis, R., Frank, K., & Tran, T. (1993). *Reducing Friction Losses in Monolithic and Segmental Bridge Tendons.* Austin: Texas Department of Transportation.

British Standards Institution. (1990). *BS5400-4 Steel, concrete and Composite Bridges.* Board of BSI.

Chen, W.-F., & Duan, L. (2000). *Bridge Engineering Handbook.* New York: CRC Press.

Committee of Land Transport Officials. (1998). *Standard Specifications for Road and Bridge Works for State Road Authorities.* Halfway House: South African Institue of Civil Engineers.

Committee of State Road Authorities. (1981). *TMH7 Part 1 and 2 - Code of Practice for the Design of Highway Bridgesand Culverts in South Africa.* Pretoria: CSRA.

Committee of State Road Authorities. (1989). *TMH7 Part 3 - Code of Practice for the Design of Highway Bridges and Culverts in South Africa.* Pretoria: CSRA.

Corven, J., & Moreton, A. (2013). *Post-Tensioning Tendon Installation and Grouting Manual.* Washington: U.S. Department of Transportation - Federal Highway Administration.

European Committee for Standardization. (2005). *Eurocode 3: Design of steel structures - Part 1-1.* Brussels: BSI British Standards.

European Organisation for Technical Approvals. (2013). *ETA-13/0815.* Vienna: Austria Institute of Construction Engineering.

Haas, T. (2014). Are Reinforced Concrete Girder Bridges More Economical Than Structural Steel Girder Bridges? A South African Perspective. *Jordan Journal of Civil Engineering, Volume 8, No. 1*, 15.

Hewson, N. (2012). *Prestressed Concrete Bridges.* London: Thomas Telford.

Huisman, H. (1972, APril). Across the Kowie River. *Civil Engineering*, p. 162.

Kreger, M. (1998). *Measured Behaviour of a Curved Precast Segmental concrete bridge erected by balanced cantilevering.* Austin: University of Texas.

Kruger, E., Newmark, A., & Smuts, M. (2008). Mountain Pass Slope Failure Retrofitted with Half Viaduct Brdge Structure, South Africa. *Structural Engineering International*, 318-322.

Leonhardt, F. (1964). *Prestressed Concrete Design and Construction.* Berlin: Wilhelm Ernst & Sohn.

Liebenberg, A. (1984). Partially Prestressed Concrete. *Concrete Beton nr 33*, 8-17.

Lin, T. (1970). *Design of Prestressed Concrete Structures.* United States of America: John Wiley & Sons, Inc.

Long, J. (1971). Some aspects of modern post-tensioning. In H. E. al., *Developments in bridge design and construction* (pp. 480-481). London: Lockwood.

Mato, F., Cornejo, M., & Sánchez, J. (2011). Design and Construction of Composite Tubular Arches with Network Suspension System: Recent Undertakings and Trends. *Journal of Civil Engineering and Architecture*, 191-214.

Naaman, A. (1982). Optimum Design of Prestressed Concrete Tension Members. *Journal of the Structural Division, ASCE*, 1722-1738.

Nawy, E. (2008). *Concrete Construction Engineernig Handbook*. Boca Raton: CRC Press.

Nielson, A. (1978). *Design of Prestressed Concrete*. Canada: John Wiley & Sons, Inc.

Nuclear Energy Agency. (1997). *Proceedings of the joint WANO/OECD-NEA Workshop Prestress Loss in NPP Containments*. Paris: Organisation for Economic Co-operation and Development (OECD).

Parke, G., & Hewson, N. (2008). *ICE manual of bridge engineering*. London: Thomas Telford.

Reis, A., & Pedro, J. J. (2019). *Bridge Design: Concepts and Analysis*. Glasgow: John Wiley and Sons.

Robberts, J. M., & Marshall, V. (2000). *Prestressed Concrete Design and Practice*. Pretoria: Concrete Society of Southern Africa Prestressed Concrete Division.

Ronné, P., Newmark, A., du Toit, N., & van Wijk, a. H. (2018). New Ashton Arch - functional assessment of direct and indirect construction costs and evaluation of service life with respect to flood risk. *International Conference on Concrete Repair, Rehabilitation and Retrofitting (ICCRRR 2018)* (p. 6). Cape Town: MATEC Web of Conferences.

Scordelis, A. (1984). Computer Models for Nonlinear Analysis of Reinforced & Presetressed Concrete Structures. *PCI Journal*, 116-135.

Smuts, M. (2018). *Guidelines for Bridge Design*. Bellville: AECOM.

South African National Roads Authority Limited. (2011). *Construction Monitoring Manual for Bridges and Structures*. Pretoria: The South African National Roads Agency Ltd.

Stavridis, L. (2010). *Structural Systems: behaviour and design volume 1*. London: Thomas Telford Limited.

Weaver, C. (2011, Fall). Re-Creating History: Modern Techniques Preserve Character of Historic Bridge. *ASPIRE*, pp. 34-36.

Wheen, R. (1979). Prestressed Concrete Members in Direct Tension. *Journal of the structural ASCE*, 1471-1487.

Wiessler, H. (1984). State of the Art. *Concrete Beton Nr. 33*, 3-6.

Xanthakos, P. (1994). *Theory and design of bridges*. New York: John Wiley & Sons, Inc.

Zong, Z., Lin, X., & Niu, J. (2015). Finite element model validation of bridge based on structural health monitoringd Part I: Response surface-based finite element model updating. *Journal of traffic and transportation Engineering*, 258-278.

APPENDIX A: CONCRETE TIE-MEMBER EXAMPLE

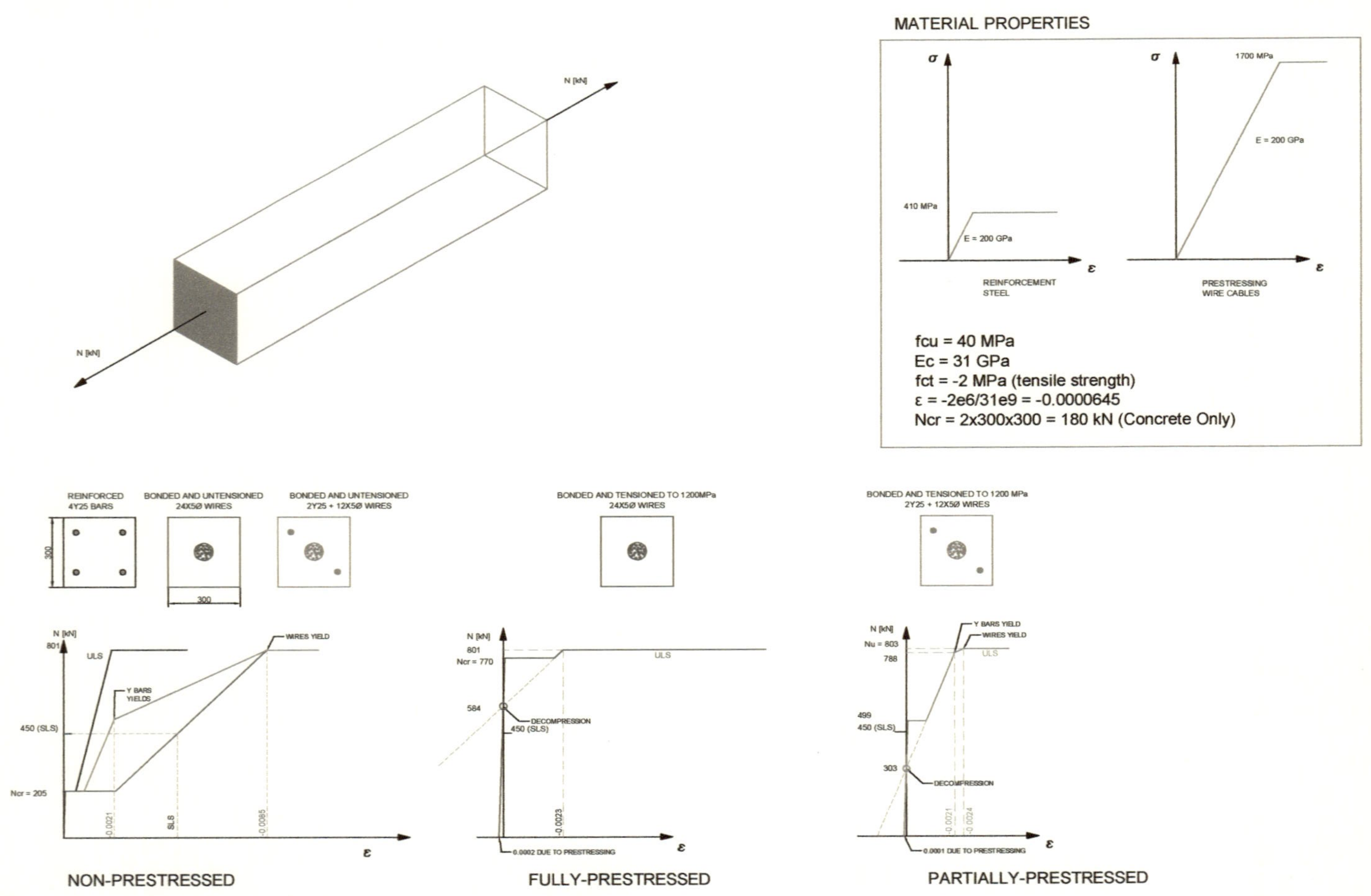

CONCRETE TIE-MEMBER: 300X300 mm CROSS SECTION

DETAIL/FIGURE No.

APPENDIX B: STRAIN GAUGE MEASUREMENT

Sequence	Element	Tendon	Jack Point	Jack Position	STRAIN Ɛ BEFORE POST-TENSIONING							STRAIN Ɛ AFTER POST-TENSIONING						
					SOUTH TIE-BEAM			NORTH TIE-BEAM				SOUTH TIE-BEAM			NORTH TIE-BEAM			
					MODULE 1 Point 1	MODULE 1 Point 3	MODULE 1 Point 5	MODULE 2 Point 1	MODULE 2 Point 2	MODULE 2 Point 3	MODULE 2 Point 5	MODULE 1 Point 1	MODULE 1 Point 3	MODULE 1 Point 5	MODULE 2 Point 1	MODULE 2 Point 2	MODULE 2 Point 3	MODULE 2 Point 5
STAGE 1																		
1	North TB	3	1	East														
2	South TB	10	1	East														
3	South TB	9	1	East														
4	North TB	4	1	East	-0.000086	-0.000034	0.000058	-0.000065	0.000038	-0.000112	0.000068	-0.000083	-0.000077	0.000031	-0.000085	0.000001	-0.000112	-0.000007
5	North TB	3	2	West	-0.000097	-0.000050	0.000018	-0.000092	0.000018	-0.000094	0.000001	-0.000079	-0.000088	0.000001	-0.000088	-0.000004	-0.000082	-0.000021
6	South TB	9	2	West	-0.000070	-0.000094	0.000004	-0.000084	-0.000004	-0.000076	-0.000020	-0.000070	-0.000097	0.000003	-0.000086	-0.000009	-0.000076	-0.000022
7	North TB	4	2	West	-0.000073	-0.000062	0.000041	-0.000083	0.000025	-0.000068	0.000027	-0.000072	-0.000088	0.000008	-0.000085	-0.000003	-0.000069	-0.000008
8	South TB	10	2	West	-0.000073	-0.000089	0.000007	-0.000085	-0.000004	-0.000069	-0.000009	-0.000071	-0.000088	0.000001	-0.000084	-0.000005	-0.000068	-0.000018
STAGE 2																		
9	North TB	2	1	West														
10	South TB	7	1	West														
11	North TB	5	1	West														
12	South TB	12	1	West														
13	North TB	2	2	East	-0.000166	-0.000033	0.000029	-0.000160	0.000069	-0.000230	0.000069	-0.000186	-0.000111	-0.000022	-0.000201	-0.000003	-0.000243	-0.000031
14	South TB	7	2	East	-0.000177	-0.000109	-0.000019	-0.000196	0.000001	-0.000239	-0.000027	-0.000201	-0.000128	-0.000018	-0.000195	-0.000003	-0.000235	-0.000041
15	South TB	12	2	East	-0.000189	-0.000069	0.000038	-0.000186	0.000057	-0.000238	0.000071	-0.000200	-0.000135	-0.000010	-0.000193	0.000004	-0.000227	-0.000019
16	North TB	5	2	East	-0.000199	-0.000135	-0.000011	-0.000193	0.000004	-0.000229	-0.000019	-0.000201	-0.000142	-0.000018	-0.000203	-0.000005	-0.000227	-0.000037
STAGE 3																		
17	South TB	8	1	East	-0.000205	-0.000144	-0.000020	-0.000206	-0.000008	-0.000241	-0.000040	-0.000266	-0.000164	-0.000010	-0.000205	-0.000008	-0.000238	-0.000041
18	North TB	1	1	East	-0.000266	-0.000164	-0.000010	-0.000205	-0.000008	-0.000238	-0.000041	-0.000262	-0.000163	-0.000007	-0.000259	-0.000010	-0.000282	-0.000041
19	South TB	11	1	East	-0.000263	-0.000162	-0.000008	-0.000260	-0.000011	-0.000282	-0.000042	-0.000311	-0.000175	-0.000038	-0.000265	-0.000011	-0.000287	-0.000049
20	North TB	6	1	East	-0.000335	-0.000036	-0.000007	-0.000272	0.000056	-0.000301	0.000057	-0.000328	-0.000099	-0.000035	-0.000303	0.000004	-0.000320	-0.000022
21	North TB	1	2	West	-0.000310	-0.000125	-0.000042	-0.000301	-0.000003	-0.000309	-0.000036	-0.000309	-0.000136	-0.000042	-0.000301	-0.000007	-0.000306	-0.000040
22	South TB	8	2	West	-0.000309	-0.000063	0.000014	-0.000300	0.000059	-0.000302	0.000061	-0.000300	-0.000117	-0.000026	-0.000297	0.000003	-0.000291	-0.000012
23	South TB	11	2	West	-0.000310	-0.000120	-0.000030	-0.000304	-0.000003	-0.000296	-0.000017	-0.000307	-0.000124	-0.000040	-0.000302	-0.000008	-0.000292	-0.000035
24	North TB	6	2	West	-0.000308	-0.000125	-0.000041	-0.000303	-0.000009	-0.000293	-0.000036	-0.000312	-0.000121	-0.000046	-0.000302	-0.000012	-0.000294	-0.000044

TENSIONING TENDON 4 FROM EASTERN SIDE

South Tie Beam

North Tie Beam

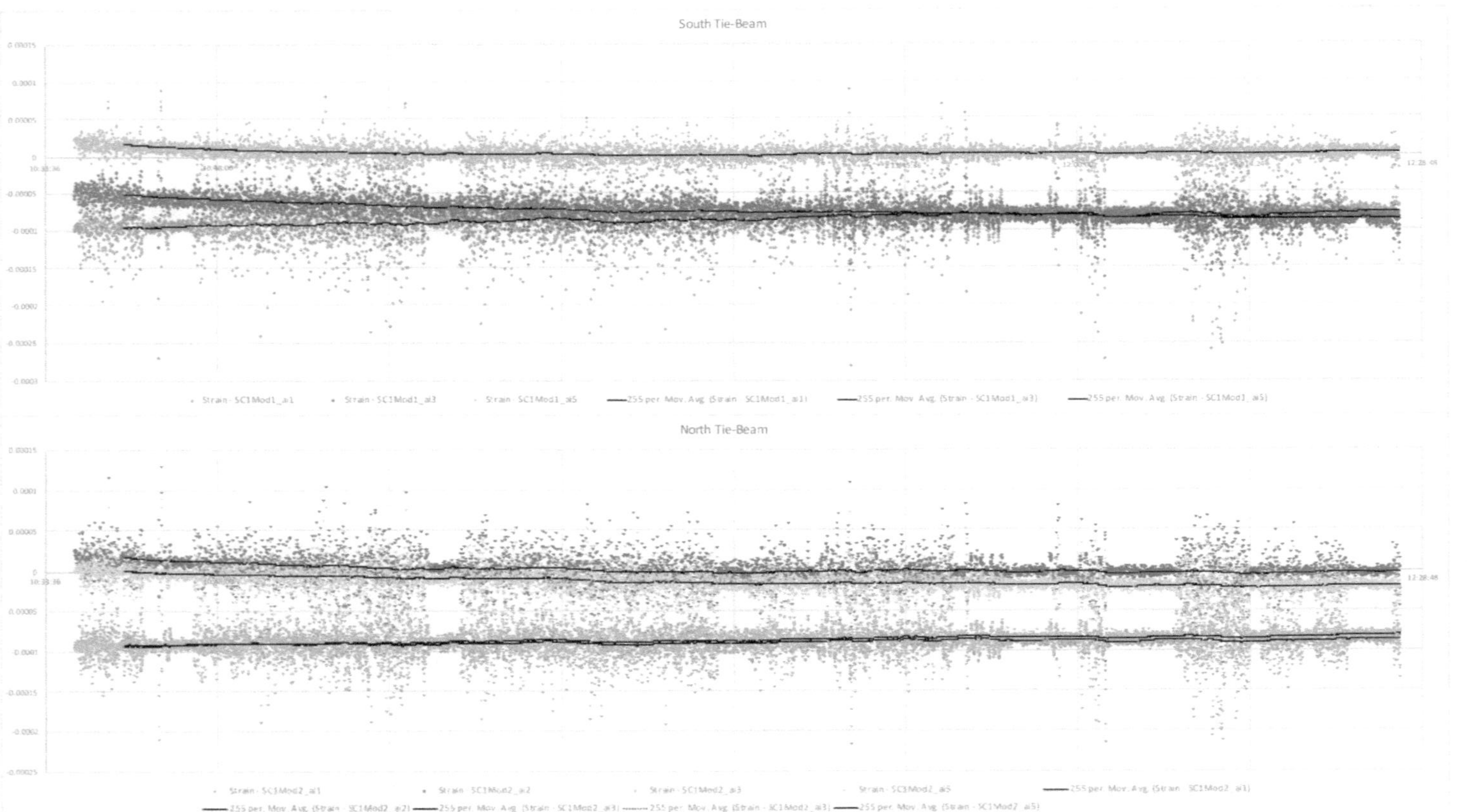

STRAIN VS TIME
TENSIONING TENDON 3 FROM WESTERN SIDE
South Tie-Beam
Strain - SC1Mod1_a1
Strain - SC1Mod1_a3
Strain - SC1Mod1_a5
255 per. Mov. Avg. (Strain - SC1Mod1_a1)
255 per. Mov. Avg. (Strain - SC1Mod1_a3)
255 per. Mov. Avg. (Strain - SC1Mod1_a5)
North Tie-Beam
Strain - SC3Mod2_a1
Strain - SC1Mod2_a2
Strain - SC1Mod2_a3
Strain - SC3Mod2_a5
255 per. Mov. Avg. (Strain - SC1Mod2_a1)
255 per. Mov. Avg. (Strain - SC1Mod2_a2)
255 per. Mov. Avg. (Strain - SC1Mod2_a3)
255 per. Mov. Avg. (Strain - SC1Mod2_a5)

TENSIONING TENDON 9 FROM WESTERN SIDE

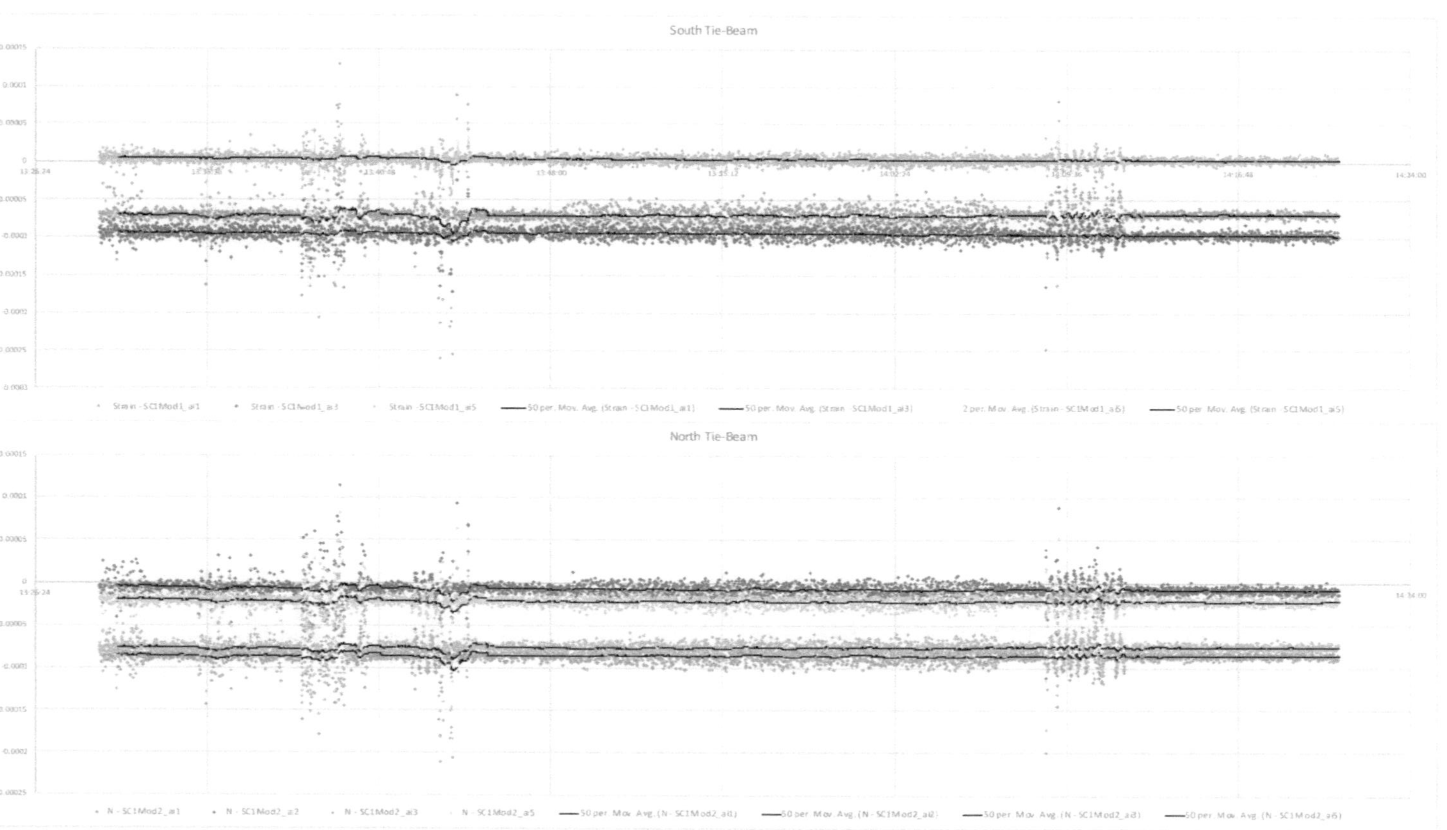

TENSIONING TENDON 4 FROM WESTERN SIDE

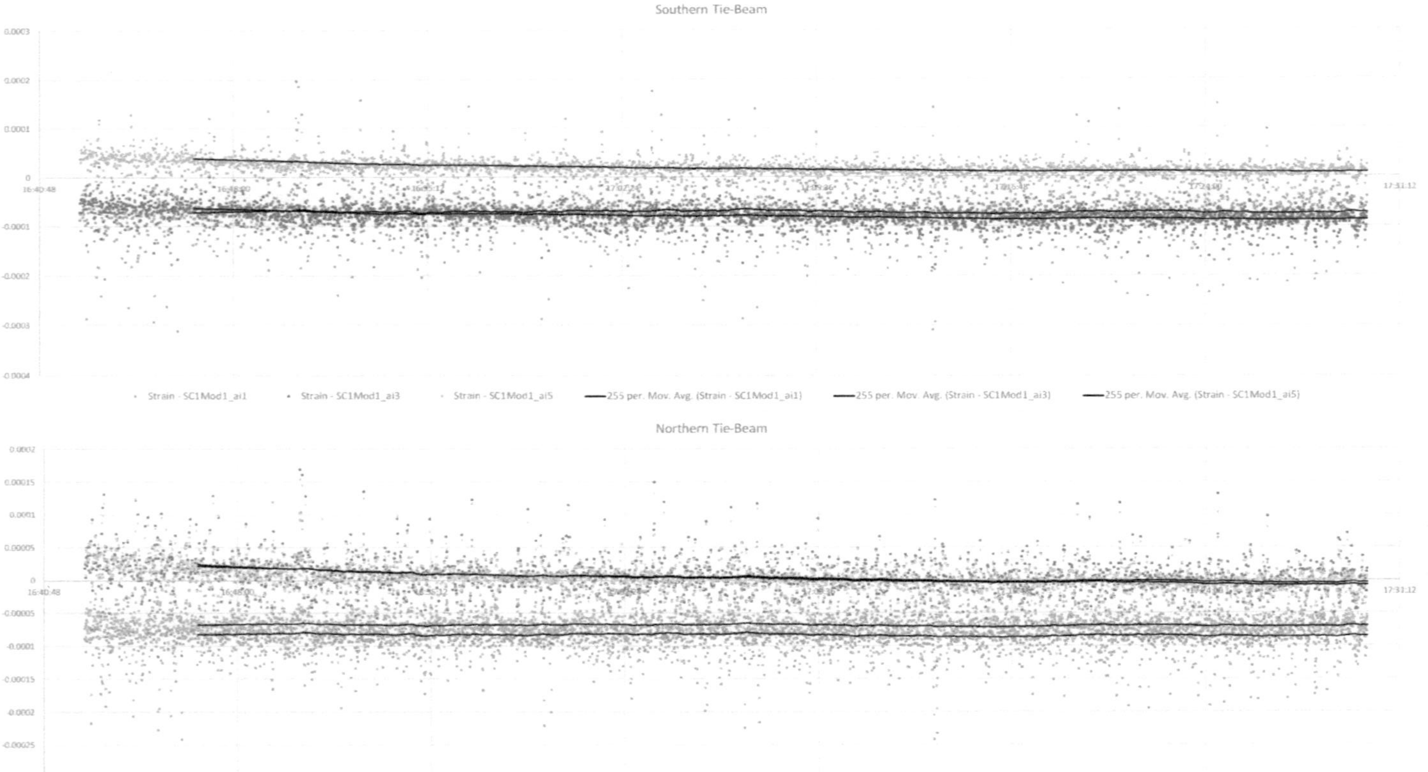

TENSIONING TENDON 10 FROM WESTERN SIDE

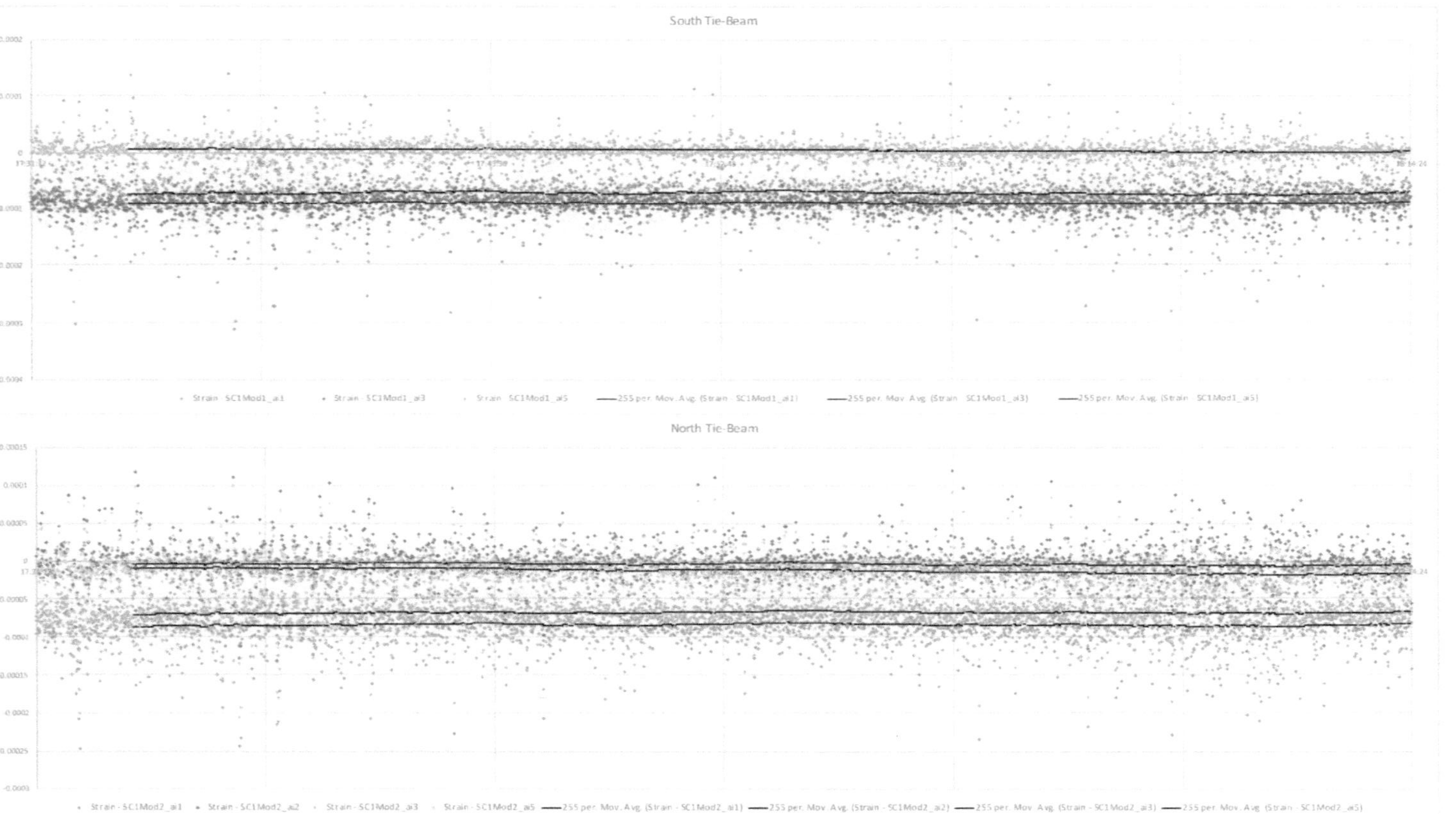

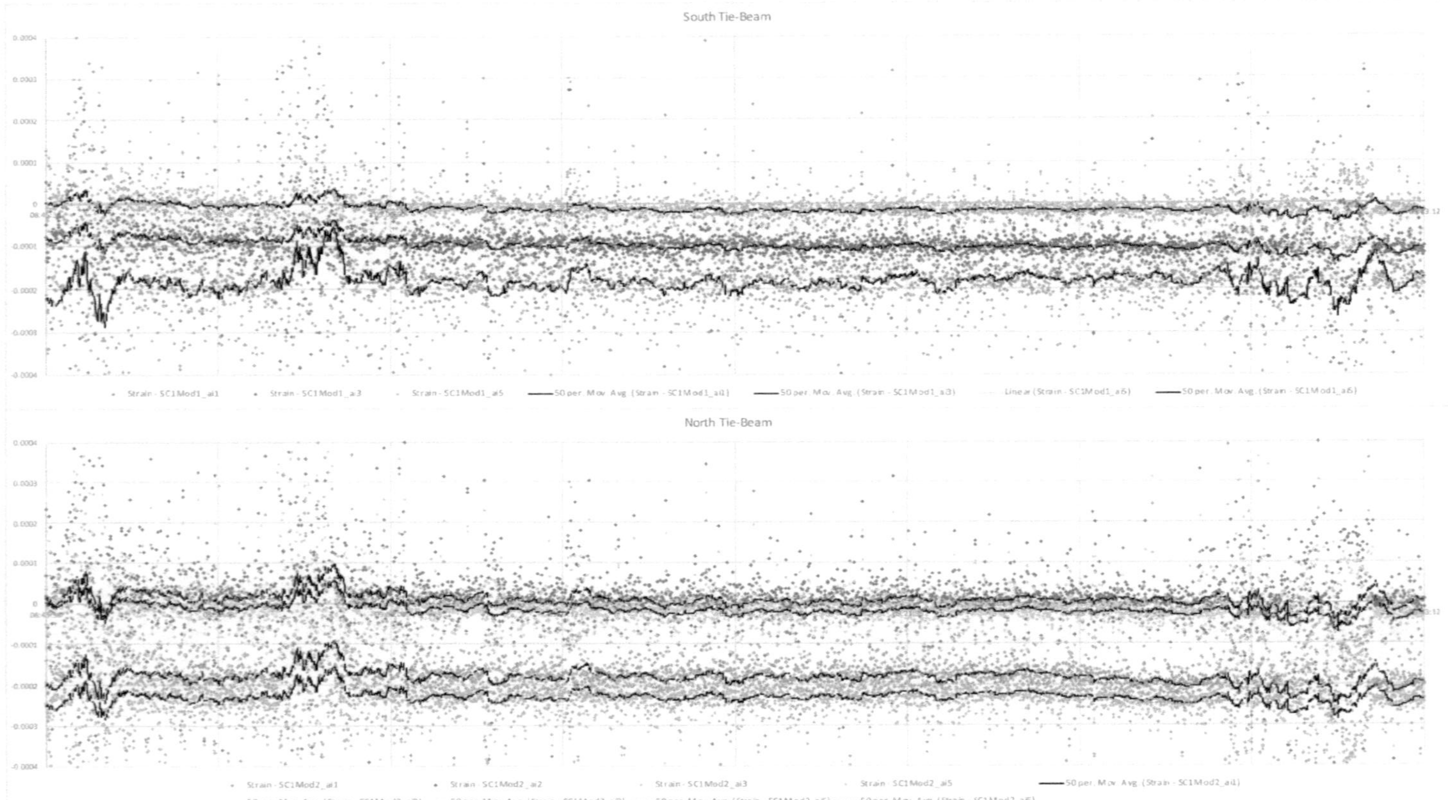

STRAIN VS TIME
TENSIONING TENDON 2 FROM EASTERN SIDE
South Tie-Beam
Strain - SC1Mod1_ai1
Strain - SC1Mod1_ai3
Strain - SC1Mod1_ai5
50 per. Mov. Avg. (Strain - SC1Mod1_ai1)
50 per. Mov. Avg. (Strain - SC1Mod1_ai3)
Linear (Strain - SC1Mod1_ai6)
50 per. Mov. Avg. (Strain - SC1Mod1_ai6)
North Tie-Beam
Strain - SC1Mod2_ai1
Strain - SC1Mod2_ai2
Strain - SC1Mod2_ai3
Strain - SC1Mod2_ai5
50 per. Mov. Avg. (Strain - SC1Mod2_ai4)
50 per. Mov. Avg. (Strain - SC1Mod2_ai2)
50 per. Mov. Avg. (Strain - SC1Mod2_ai3)
50 per. Mov. Avg. (Strain - SC1Mod2_ai6)
50 per. Mov. Avg. (Strain - SC1Mod2_ai6)

South Tie-Beam

North Tie-Beam

TENSIONING TENDON 12 FROM EASTERN SIDE

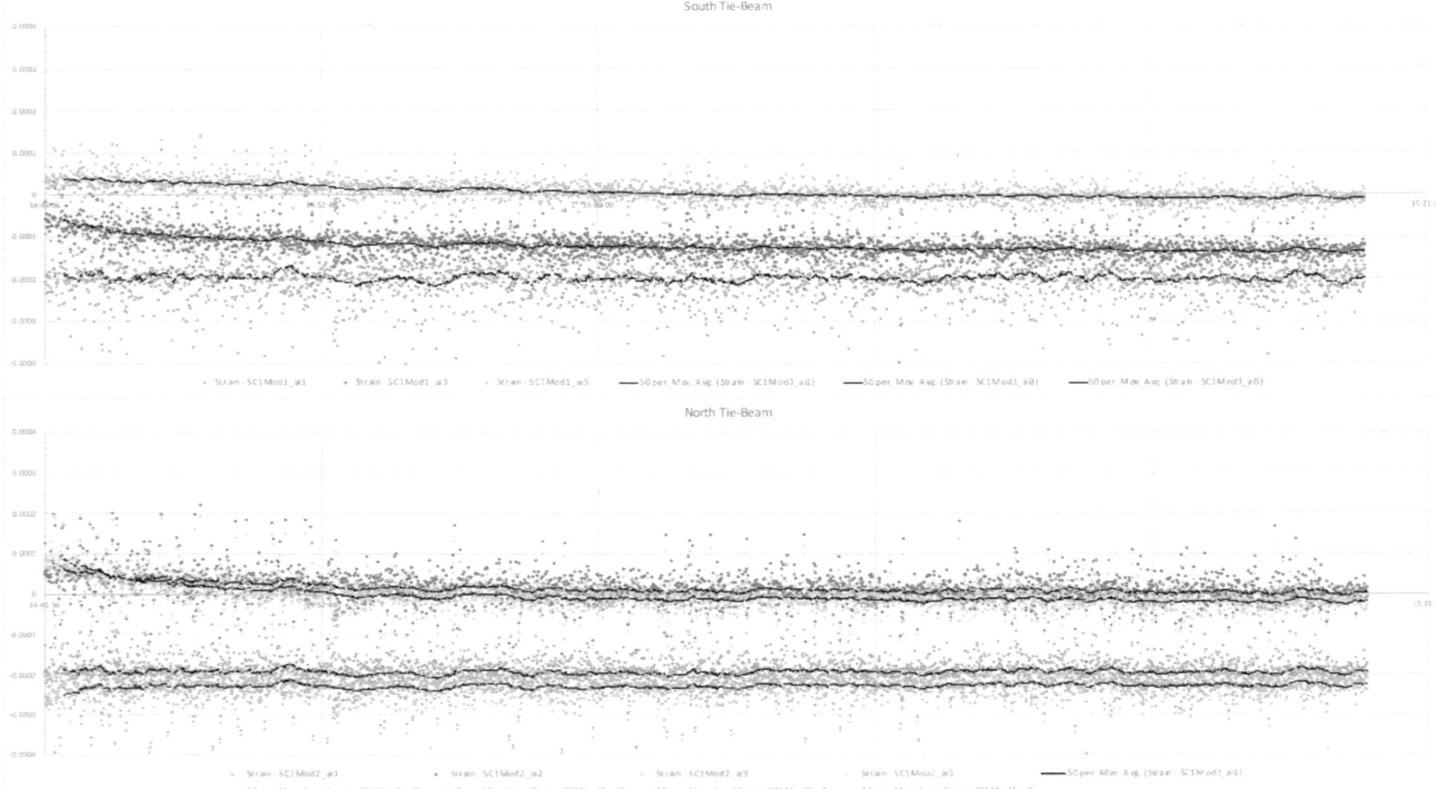

TENSIONING TENDON 5 FROM EASTERN SIDE

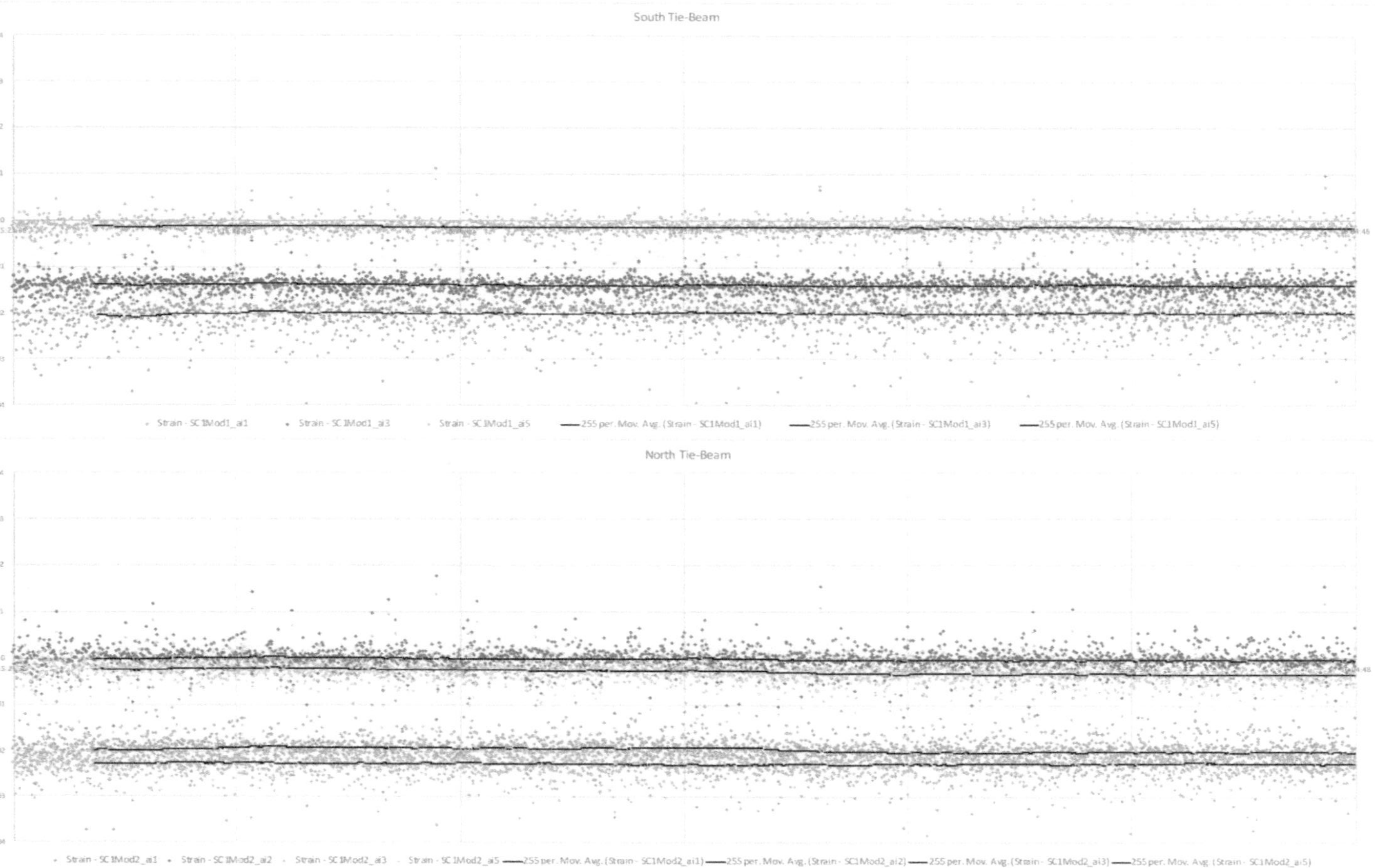

TENSIONING TENDON 8 FROM EASTERN SIDE

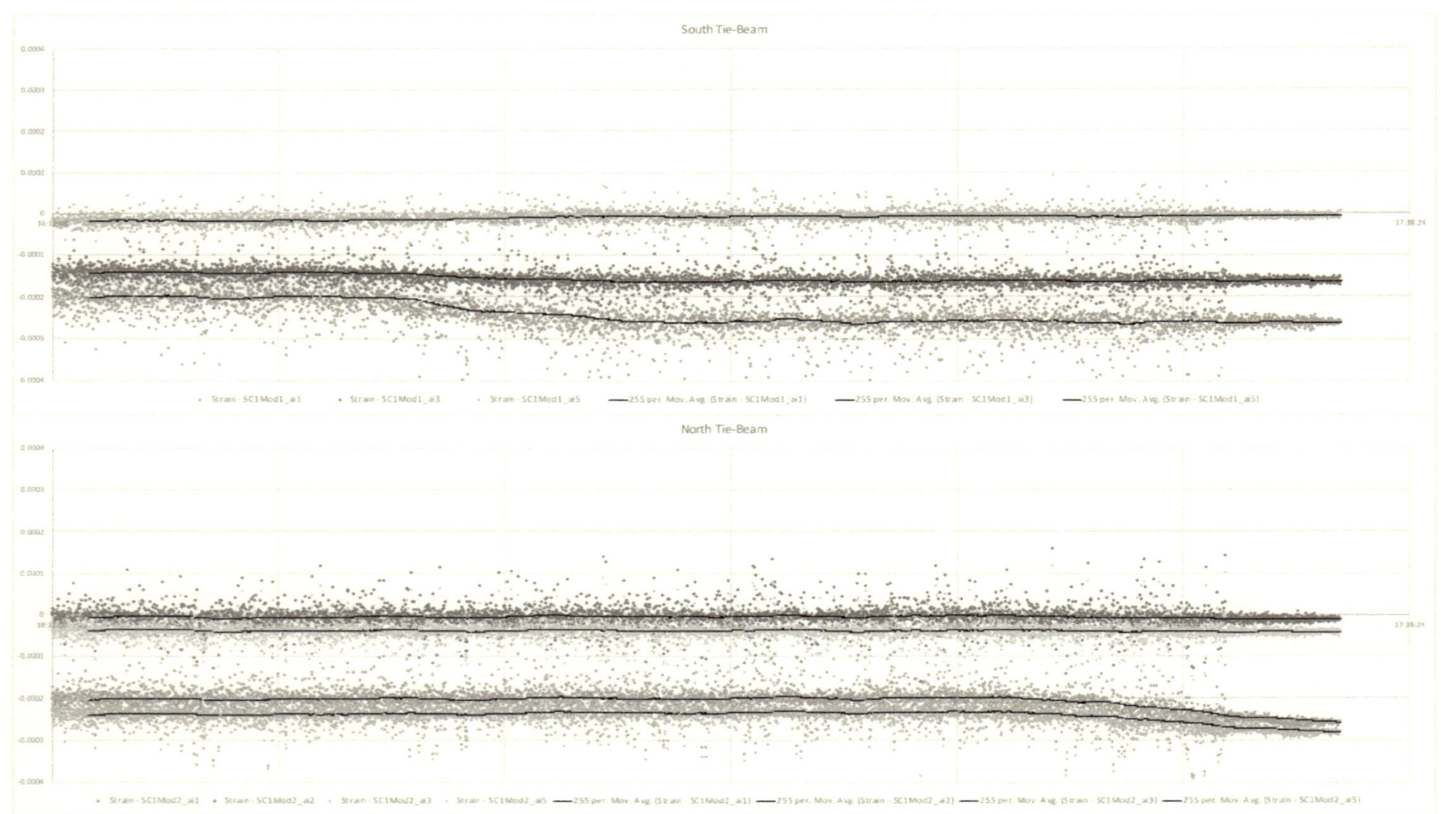

TENSIONING TENDON 1 AND TENDON 11 FROM EASTERN SIDE

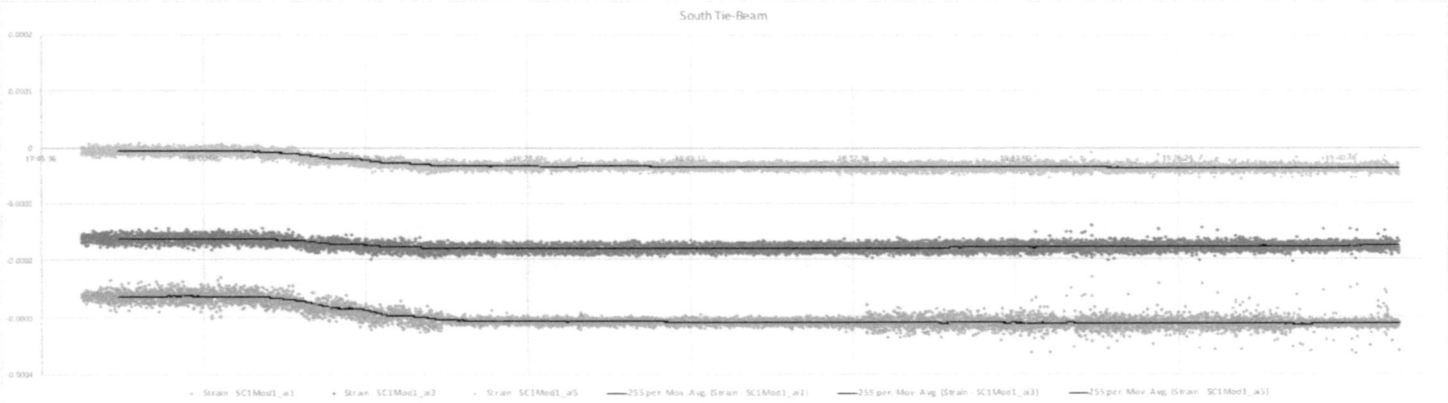

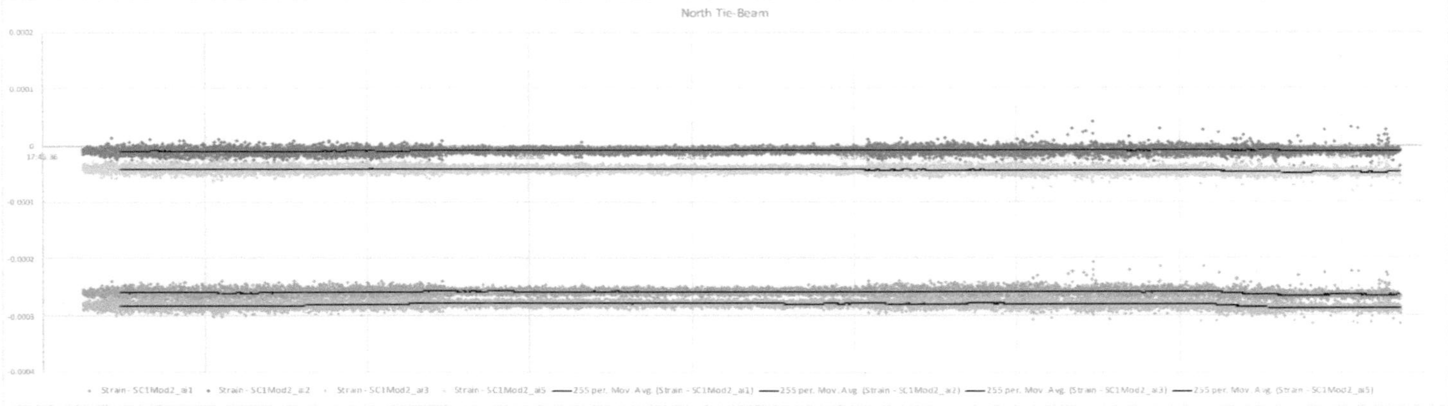

South Tie-Beam

North Tie-Beam

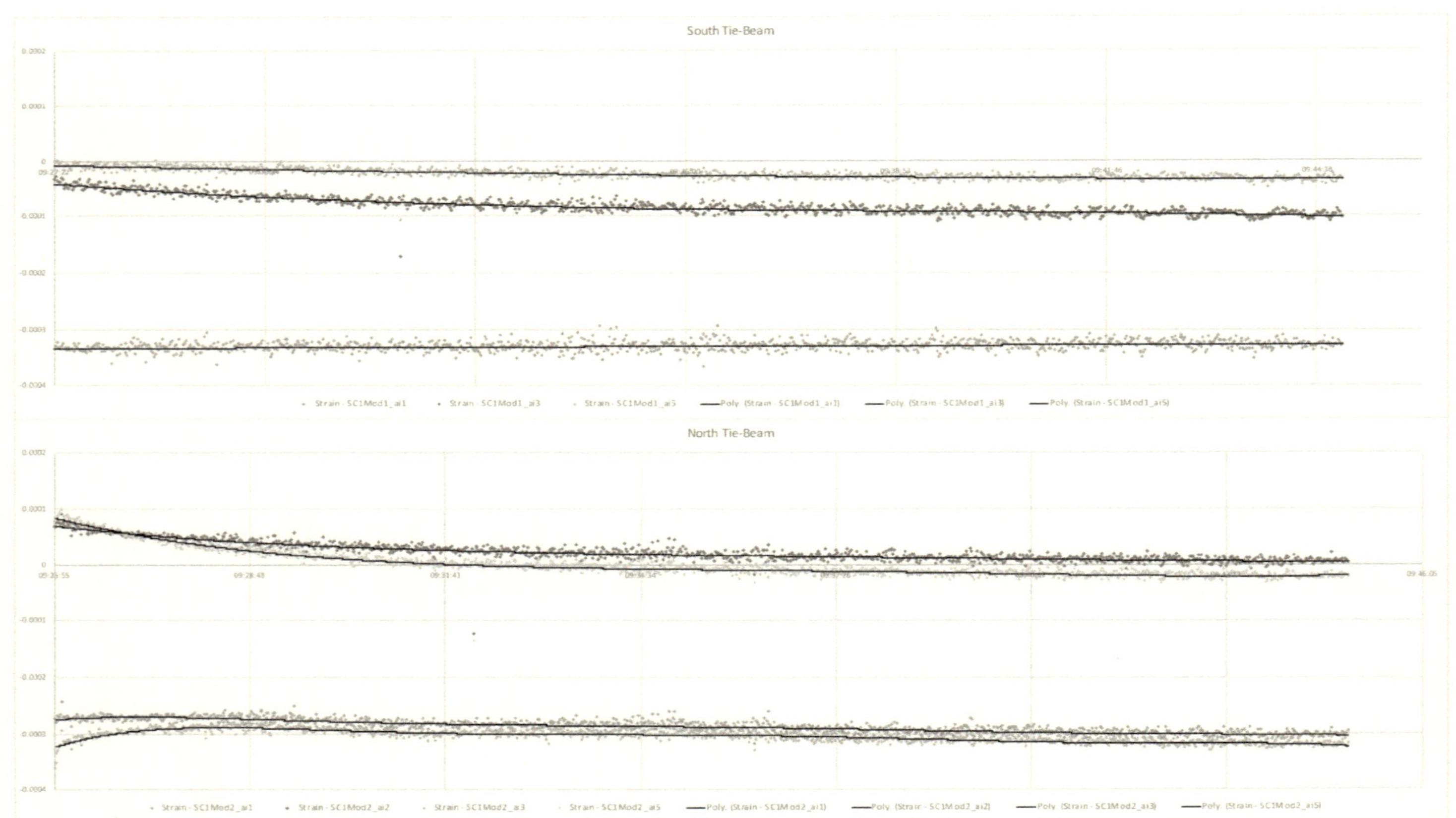

STRAIN VS TIME
TENSIONING TENDON 6 FROM EASTERN SIDE
South Tie-Beam
Strain - SC1Mod1_ai1
Strain - SC1Mod1_ai3
Strain - SC1Mod1_ai5
Poly. (Strain - SC1Mod1_ai1)
Poly. (Strain - SC1Mod1_ai3)
Poly. (Strain - SC1Mod1_ai5)
North Tie-Beam
Strain - SC1Mod2_ai1
Strain - SC1Mod2_ai2
Strain - SC1Mod2_ai3
Strain - SC1Mod2_ai5
Poly. (Strain - SC1Mod2_ai1)
Poly. (Strain - SC1Mod2_ai2)
Poly. (Strain - SC1Mod2_ai3)
Poly. (Strain - SC1Mod2_ai5)

TENSIONING TENDON 1 FROM WESTERN SIDE
South Tie-Beam
Strain - SC1Mod1_ai1
Strain - SC1Mod1_ai3
Strain - SC1Mod1_ai5
255 per. Mov. Avg. (Strain - SC1Mod1_ai1)
255 per. Mov. Avg. (Strain - SC1Mod1_ai3)
255 per. Mov. Avg. (Strain - SC1Mod1_ai5)
North Tie-Beam
Strain - SC1Mod2_ai1
Strain - SC1Mod2_ai2
Strain - SC1Mod2_ai3
Strain - SC1Mod2_ai5
255 per. Mov. Avg. (Strain - SC1Mod2_ai1)
255 per. Mov. Avg. (Strain - SC1Mod2_ai2)
255 per. Mov. Avg. (Strain - SC1Mod2_ai3)
255 per. Mov. Avg. (Strain - SC1Mod2_ai5)

STRAIN VS TIME
TENSIONING TENDON 8 FROM WESTERN SIDE

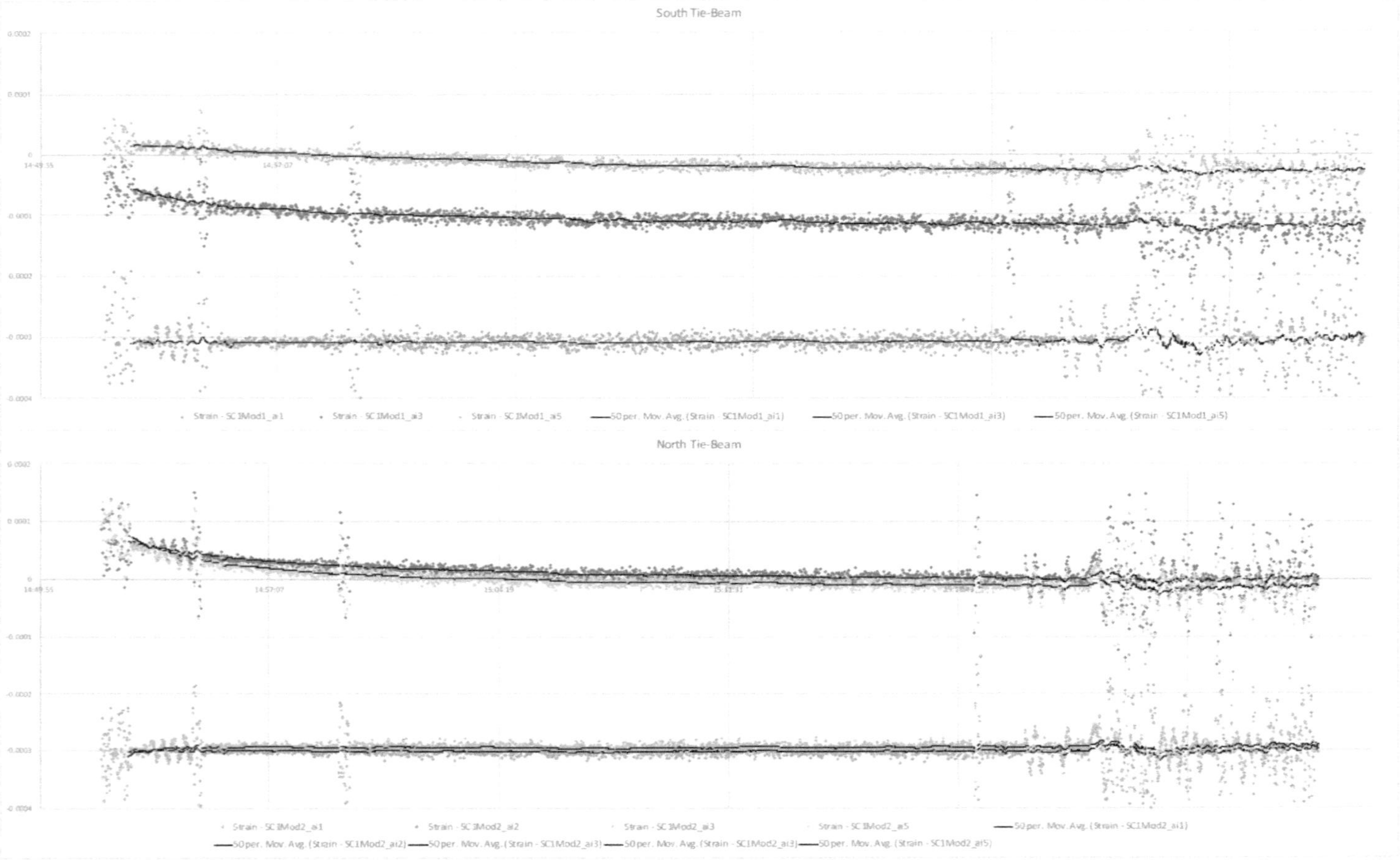

TENSIONING TENDON 11 FROM WESTERN SIDE

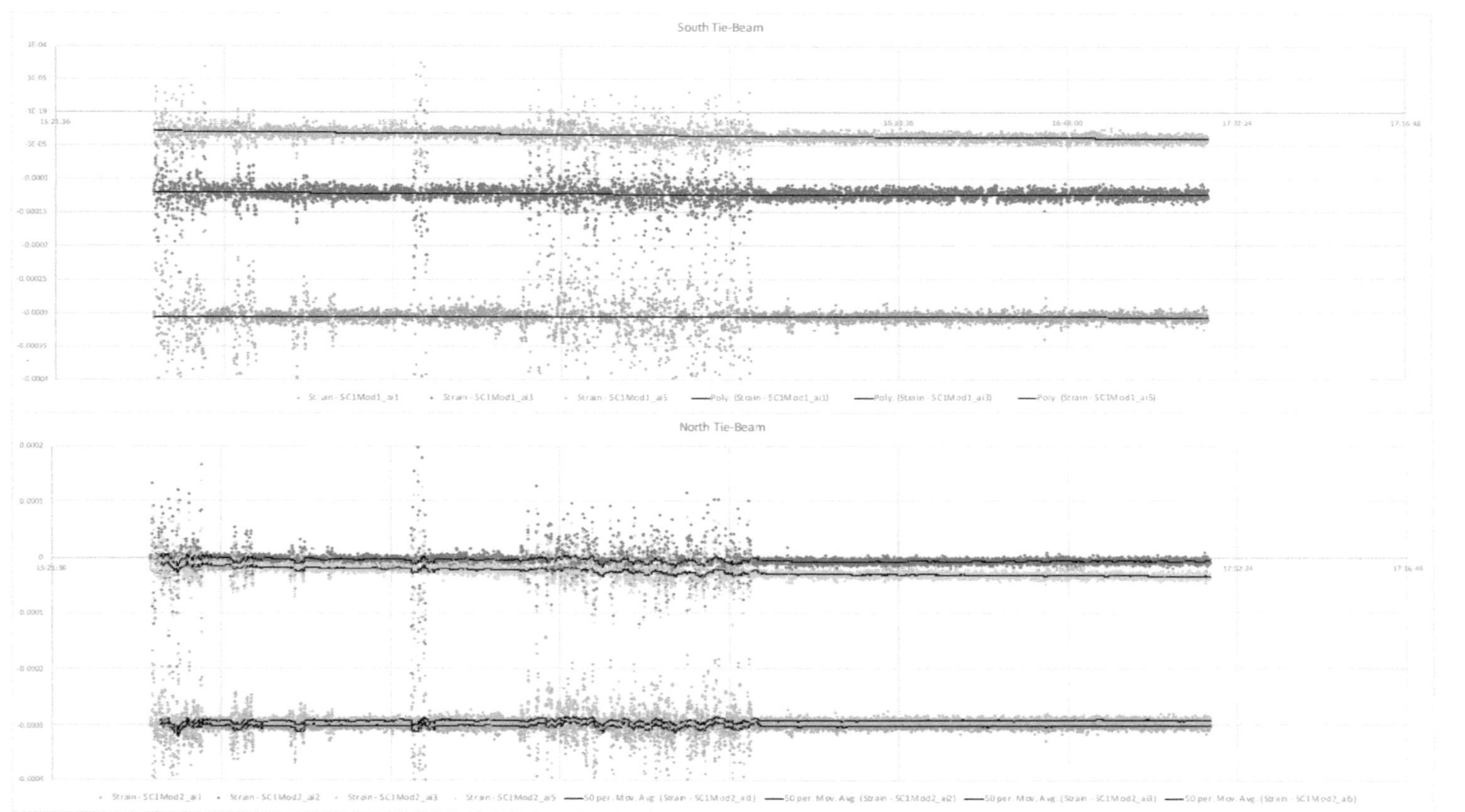

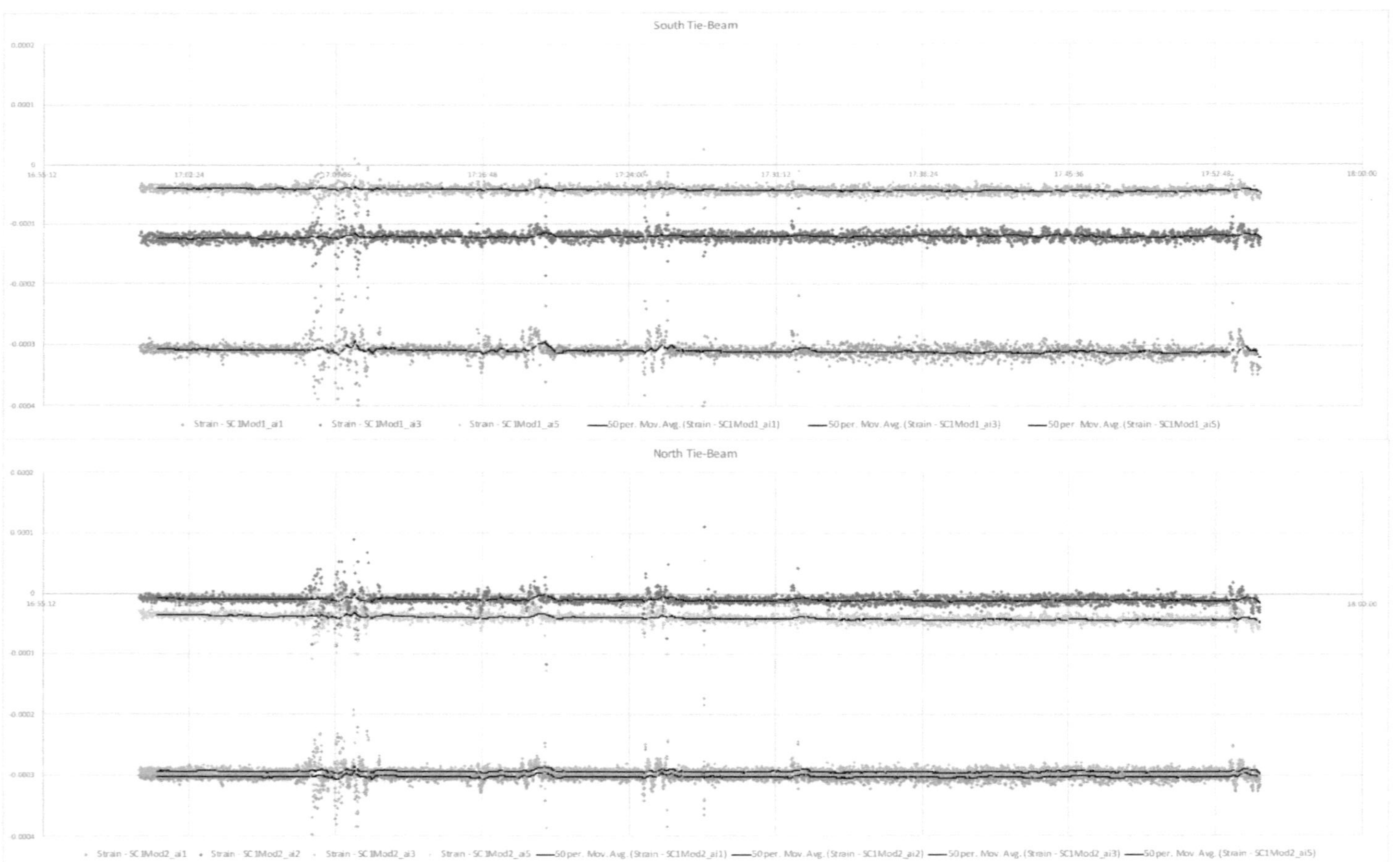
STRAIN VS TIME
TENSIONING TENDON 6 FROM WESTERN SIDE
South Tie-Beam
North Tie-Beam
Strain - SC1Mod1_ai1
Strain - SC1Mod1_ai3
Strain - SC1Mod1_ai5
50 per. Mov. Avg. (Strain - SC1Mod1_ai1)
50 per. Mov. Avg. (Strain - SC1Mod1_ai3)
50 per. Mov. Avg. (Strain - SC1Mod1_ai5)
Strain - SC1Mod2_ai1
Strain - SC1Mod2_ai2
Strain - SC1Mod2_ai3
Strain - SC1Mod2_ai5
50 per. Mov. Avg. (Strain - SC1Mod2_ai1)
50 per. Mov. Avg. (Strain - SC1Mod2_ai2)
50 per. Mov. Avg. (Strain - SC1Mod2_ai3)
50 per. Mov. Avg. (Strain - SC1Mod2_ai5)

APPENDIX C: TENDON ELONGATION MEASUREMENTS

SUMMARY OF POST-TENSIONING RESULTS

| Tendon number | Tendon Extensions (mm) | | | | MEASURED DATA | | | | | | | DESIGN DATA | | | EVALUATION | |
	Jacking Point 1	Interpolated	Jacking Point 2	Total	Actual E Modulus (GPa)	Jacking Force (kN)	Number of Strands	Length of Strand in Jack (mm)	Elongation in Jack (mm)	Dead Anchor Pull-In (mm)	Final Measured Elongation	Design E Modulus (Gpa)	Design Elongation (mm)	Adjusted Design Elongation	Percentage Difference (%)	Average (%)
CALCULATIONS	A	B	C	$D = A+B+C$	E	F	G	H	$J = FH/(G*150*E)$	K	$L = D - 2*(J + K)$	E_d	M	$N = M*E/E_d$	$P = (L-N)/N$	
TB North - Tendon 01	506	123	84	713	197.6	6 487	31	1 100	7.76	2	695.1	195	711	701.6	-0.92	
TB North - Tendon 02	479	113	121	713	196.8	6 487	31	1 100	7.80	2	695.5	195	711	704.7	-1.32	
TB North - Tendon 03	477	120	78	675	199.0	6 487	31	1 100	7.71	2	657.4	195	716	701.6	-6.29	
TB North - Tendon 04	503	132	81	716	195.0	6 487	31	1 100	7.87	2	698.2	195	716	715.9	-2.47	
TB North - Tendon 05	540	121	72	733	191.5	6 487	31	1 100	8.01	2	715.2	195	704	716.4	-0.17	
TB North - Tendon 06	482	138	87	706	194.6	6 487	31	1 100	7.89	2	688.3	195	704	704.9	-2.35	-2.25
TB South - Tendon 07	434	109	157	700	199.5	6 487	31	1 100	7.69	2	682.8	195	711	695.0	-1.75	
TB South - Tendon 08	485	123	77	684	201.3	6 487	31	1 100	7.62	2	667.1	195	711	688.7	-3.13	
TB South - Tendon 09	499	123	70	692	195.0	6 487	31	1 100	7.87	2	674.1	195	716	716.0	-5.86	
TB South - Tendon 10	436	107	109	652	199.1	6 487	31	1 100	7.71	2	635.1	195	716	701.4	-9.46	
TB South - Tendon 11	523	127	42	691	202.1	6 487	31	1 100	7.59	2	673.8	195	704	678.9	-0.75	
TB South - Tendon 12	521	128	63	712	204.5	6 487	31	1 100	7.50	2	695.4	195	704	670.7	3.68	-2.88
																-2.57

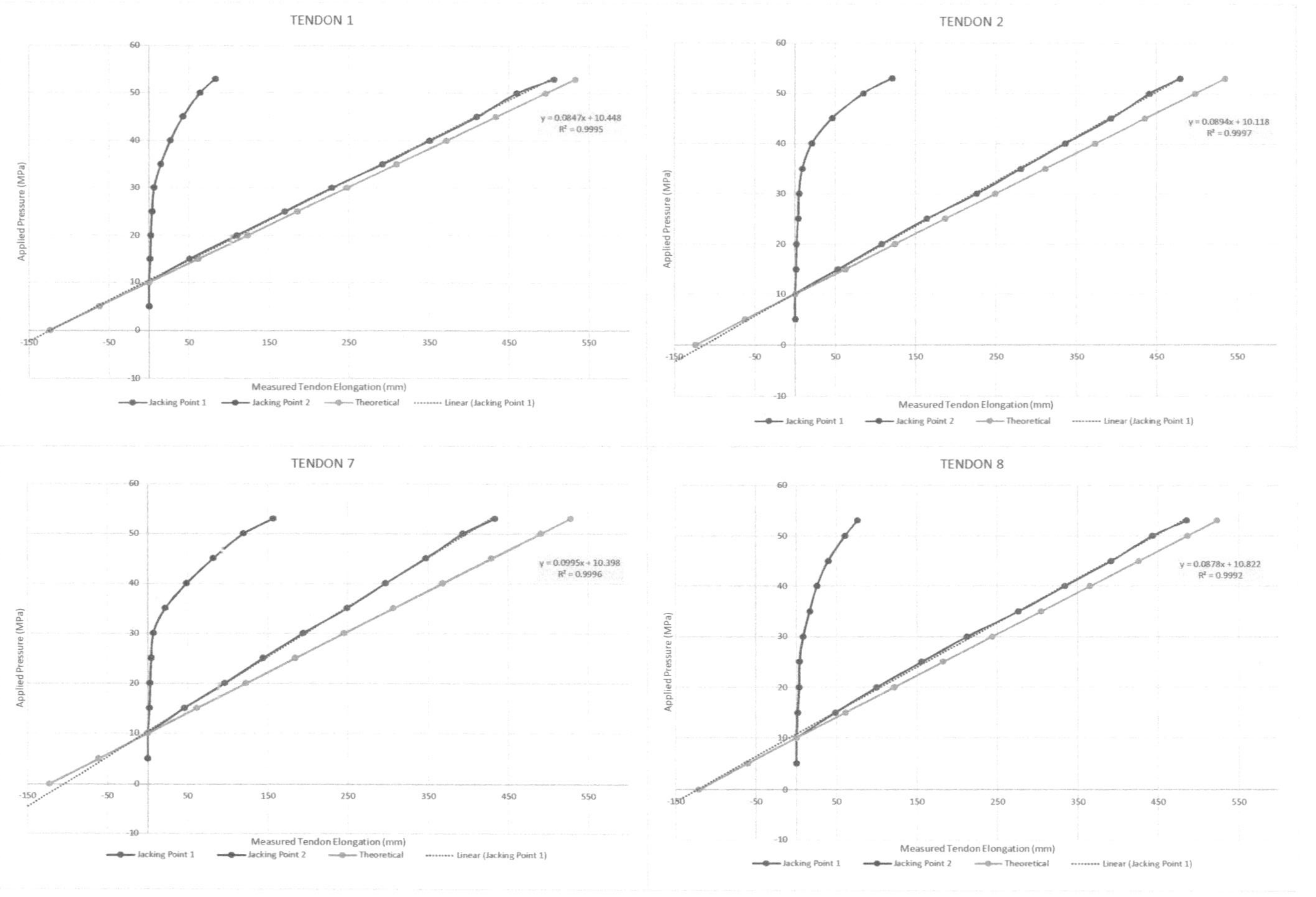
TENDON 1
y = 0.0847x + 10.448
R² = 0.9995
Applied Pressure (MPa)
Measured Tendon Elongation (mm)
Jacking Point 1
Jacking Point 2
Theoretical
Linear (Jacking Point 1)
TENDON 2
y = 0.0894x + 10.118
R² = 0.9997
Applied Pressure (MPa)
Measured Tendon Elongation (mm)
Jacking Point 1
Jacking Point 2
Theoretical
Linear (Jacking Point 1)
TENDON 7
y = 0.0995x + 10.398
R² = 0.9996
Applied Pressure (MPa)
Measured Tendon Elongation (mm)
Jacking Point 1
Jacking Point 2
Theoretical
Linear (Jacking Point 1)
TENDON 8
y = 0.0878x + 10.822
R² = 0.9992
Applied Pressure (MPa)
Measured Tendon Elongation (mm)
Jacking Point 1
Jacking Point 2
Theoretical
Linear (Jacking Point 1)

APPLIED PRESSURE vs TENDON ELONGATION

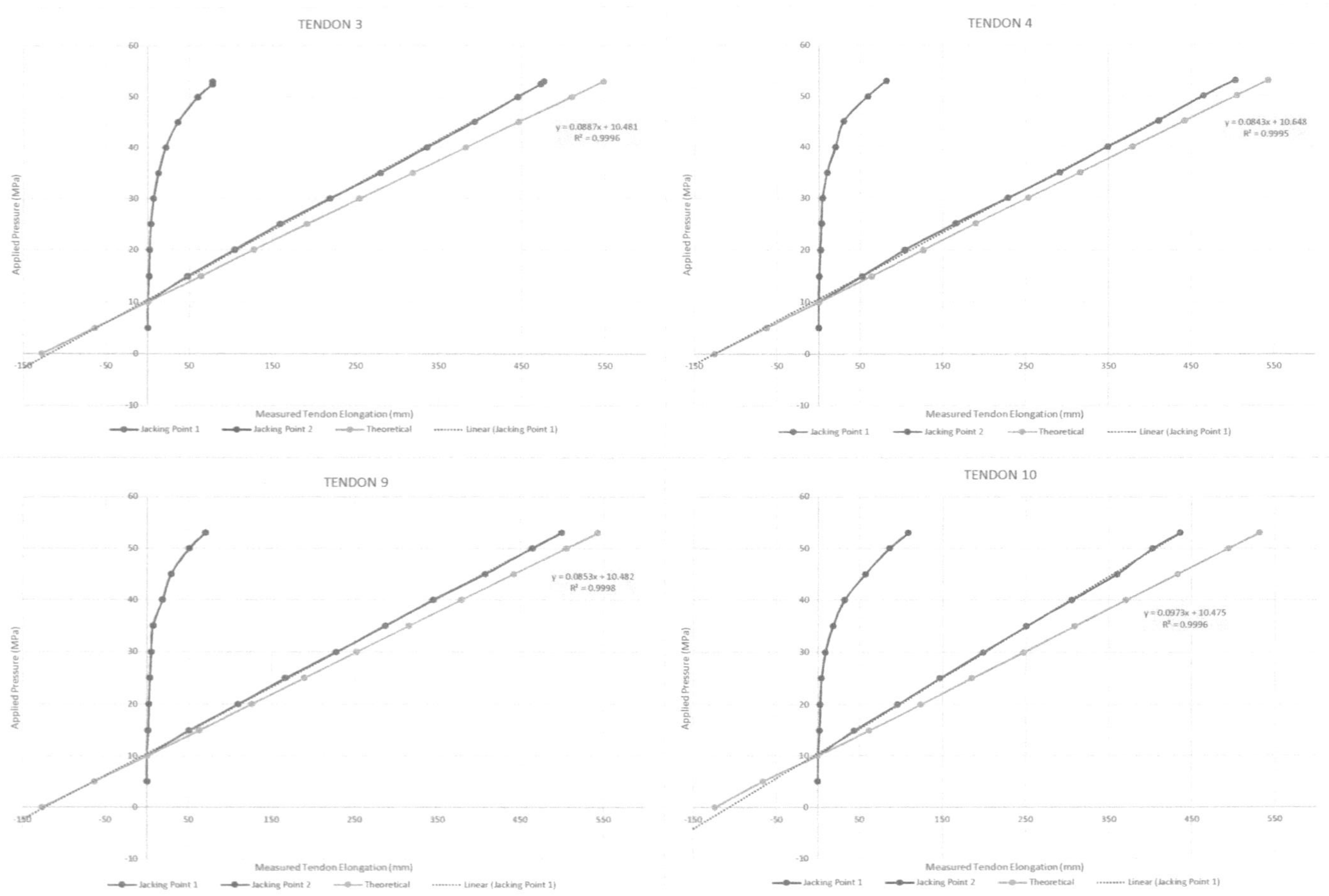

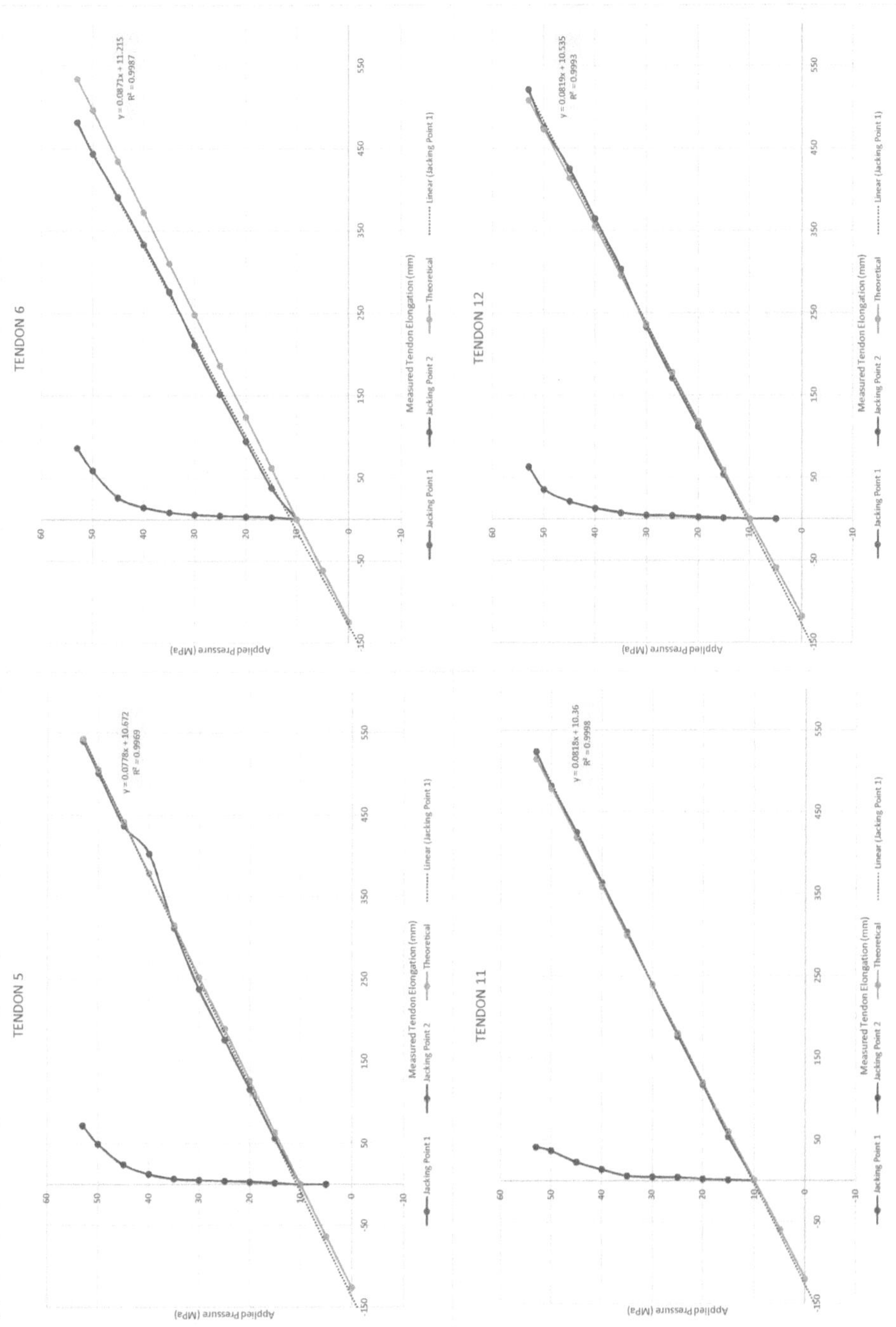

APPENDIX D: LIFT OFF TEST RESULTS

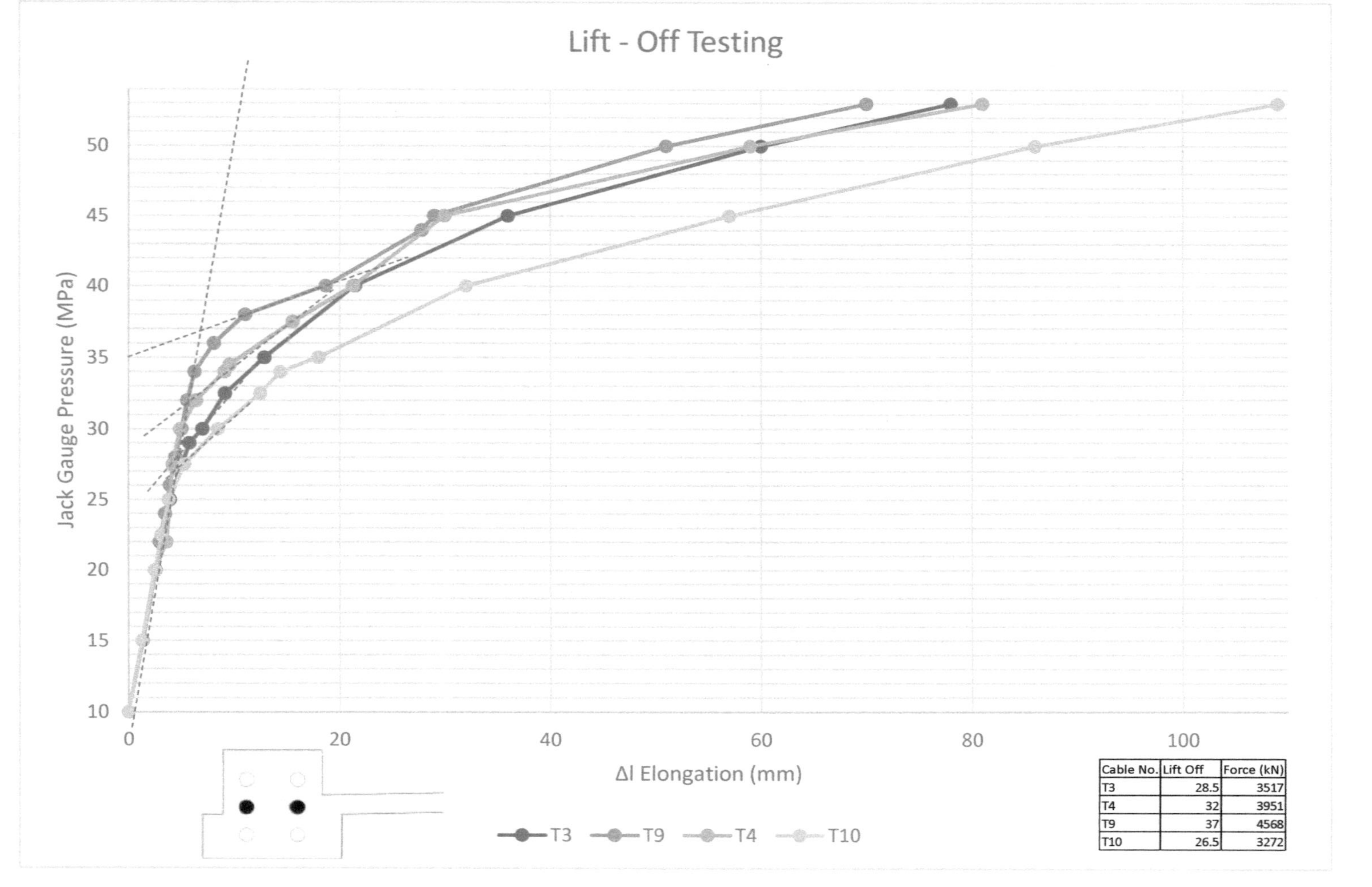

Lift - Off Testing
Jack Gauge Pressure (MPa)
Δl Elongation (mm)
T3
T9
T4
T10
Cable No. | Lift Off | Force (kN)
T3 | 28.5 | 3517
T4 | 32 | 3951
T9 | 37 | 4568
T10 | 26.5 | 3272

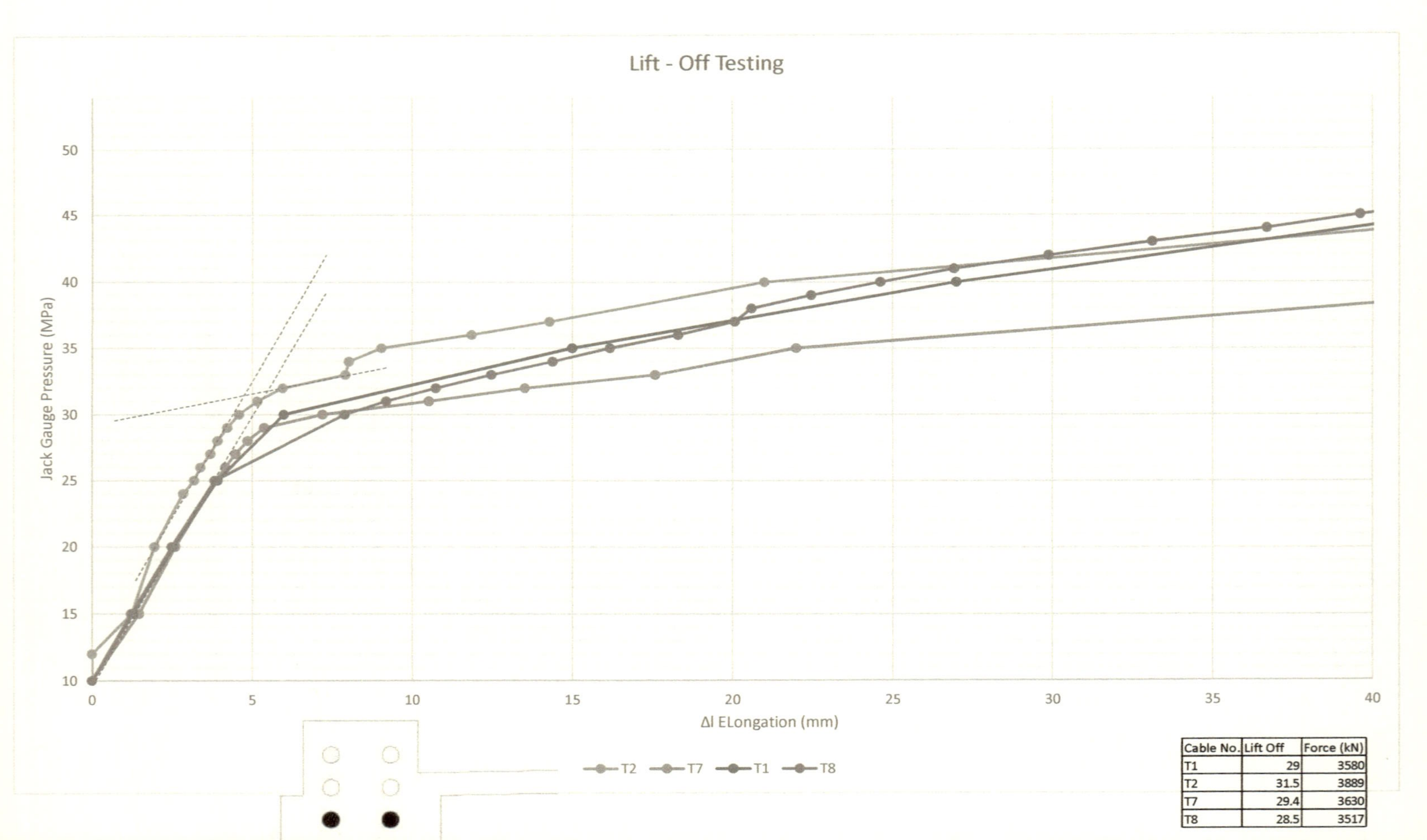

Cable No.	Lift Off	Force (kN)
T1	29	3580
T2	31.5	3889
T7	29.4	3630
T8	28.5	3517

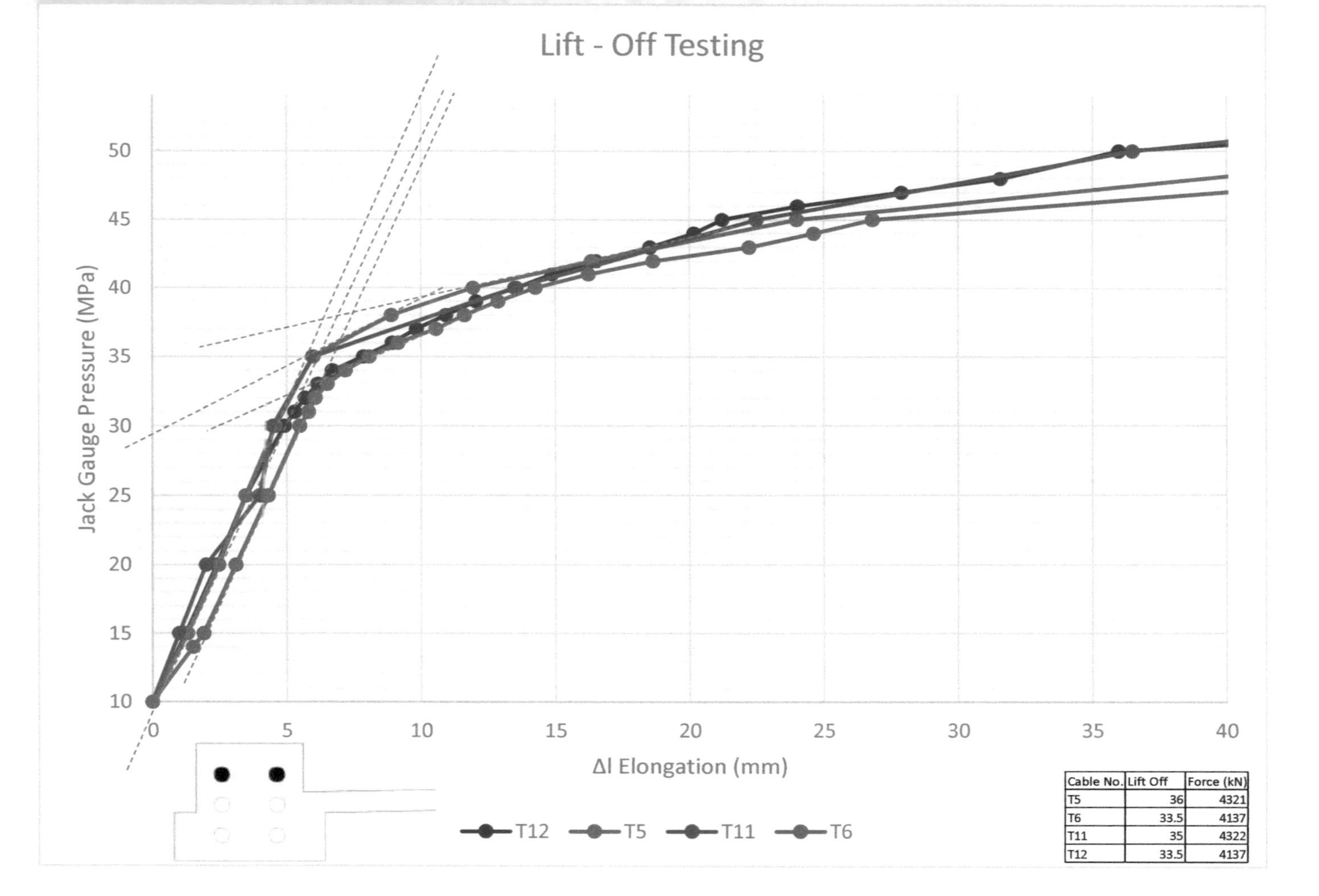

Cable No.	Lift Off	Force (kN)
T5	36	4321
T6	33.5	4137
T11	35	4322
T12	33.5	4137

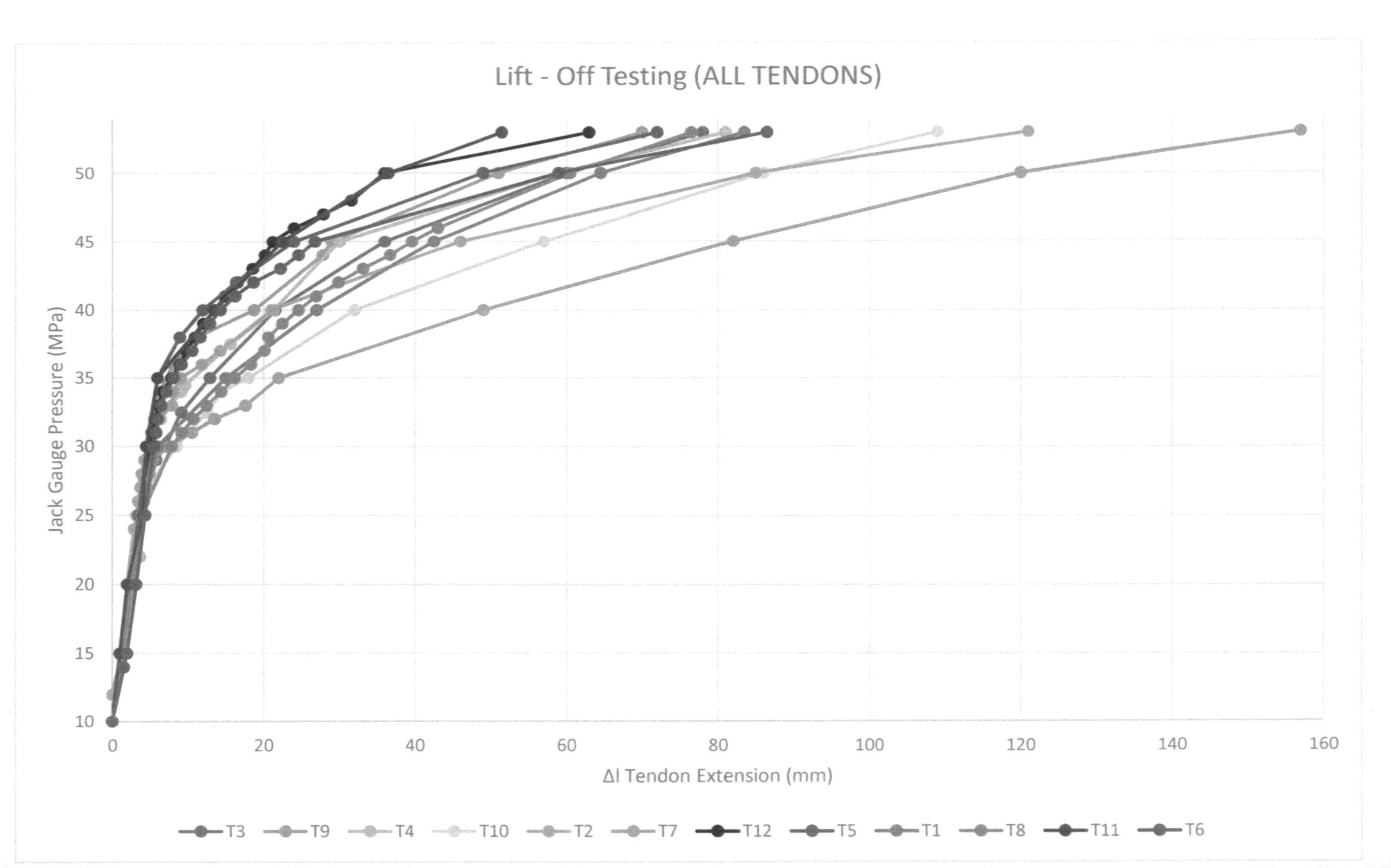

Lift - Off Testing (ALL TENDONS)
Jack Gauge Pressure (MPa)
ΔI Tendon Extension (mm)
T3
T9
T4
T10
T2
T7
T12
T5
T1
T8
T11
T6

www.ingramcontent.com/pod-product-compliance
Lightning Source LLC
LaVergne TN
LVHW041658190726
843493LV00007B/1852